Applications of Federated Learning in Technological Advancements

This book explores the applications and advancements of federated learning across diverse sectors, focusing on its integration with cutting-edge technologies like Internet of Things (IoT), artificial intelligence (AI), blockchain, and digital twins. Real-world examples and case studies illustrate federated learning's role in healthcare, smart cities, and maritime applications while addressing critical concerns such as security. It provides insights into federated learning's transformative potential, offering practical strategies for intelligent systems and sustainable environments. The book particularly:

- Focuses on the federated learning–based model optimization, addressing the significance of IoT and federated learning in the evolution of intelligent systems for various applications;
- Describes the different optimization techniques of federated learning systems from a practical point of view;
- Highlights economic, social, and environmental impacts of smart technologies and provides insights into IoT, 5G/6G communication, and computing standards;
- Provides analysis of the use cases of federated learning regarding the development of IoT, AI, blockchain, digital twins;
- Offers strategies for overcoming challenges associated with federated learning systems, including connectivity, computation, threats, privacy, and security issues.

It covers fundamental concepts, practical implementations, and trends to serve as a reference resource for professionals and researchers in the field.

Applications of Federated Learning in Technological Advancements

Use Cases and Applications

Edited by
S. Jayachitra, A. Prasanth, Rajesh Kumar Dhanaraj,
Balasubramaniam S, and Danilo Pelusi

CRC Press
Taylor & Francis Group
Boca Raton London New York

CRC Press is an imprint of the
Taylor & Francis Group, an informa business

Designed cover image: ShutterStock

First edition published 2026
by CRC Press
2385 NW Executive Center Drive, Suite 320, Boca Raton FL 33431

and by CRC Press
4 Park Square, Milton Park, Abingdon, Oxon, OX14 4RN

CRC Press is an imprint of Taylor & Francis Group, LLC

ISBN: 9781032859514 (hbk)
ISBN: 9781032873701 (pbk)
ISBN: 9781003532323 (ebk)

DOI: 10.1201/9781003532323

Typeset in Times
by Newgen Publishing UK

Contents

Preface

This book contributes an overview of advanced technologies used in federated learning (FL) for various applications and offers an indispensable resource for professionals and researchers in this field. The security and privacy concerns focused on implementing FL in various sectors make it a valuable resource for those who are concerned about these issues. The book provides valuable insights into the future trends and potential developments in FL for healthcare, making it an essential resource for professionals and researchers who are looking to stay ahead of the curve.

The book is written for a broad audience, including professionals, researchers, and students from different disciplines such as artificial intelligence (AI) and internet of things (IoT). The book also includes real-world examples and case studies to illustrate the concepts and applications of FL in technological advancements, making it an accessible and engaging resource for those who are new to this field. Solicit contributions of this book include theory, applications, and design schemes of intelligent systems, smart cities, blockchain, and maritime applications, as well as the methodologies behind them.

FL enables decentralized model training by allowing multiple devices or organizations to collaboratively develop a shared model without transferring their local data to a central server. It ensures that sensitive information remains on local devices, addressing privacy concerns while harnessing the power of distributed datasets. Several businesses have adopted FL integration in IoT by providing innovative, secure, and intelligent solutions across multiple domains. It involves intelligent and self-adapting vehicles that strengthen performance, safety, and production and can adjust to altering conditions naturally.

Healthcare AI applications can protect secret patient data, customize financial services, detect fraud, and manage traffic in feasible urban environments. The amalgamation of FL, AI, and IoT is pushing toward new territories of analysis and processing. These inventions cause a paradigm change in technology by redefining the relationship between humans and machines. The availability of 5G's and 6G's and the standout features that they present are improving the capacity of federated reinforcement learning in IoT. Adoption of 5G/6G modern technologies enhances the mobility, efficiency, and reliability of sharing data while lowering its latency which is crucial in the extendibility of IoT systems.

FL enables many devices or edge nodes to train a model without sharing their local data with other nodes. FL preserves the data locally in the devices it is produced and only sends model updates containing gradients or other model parameters to a central server or aggregator as opposed to traditional centralized machine learning. Blockchain integrated FL emerges as a promising solution to address the intricate challenges surrounding privacy, security, and data integrity within the context of IoT-based smart applications. It demonstrates a collaborating advantage that greatly enhances data privacy, security, and scalability.

Unmanned aerial vehicles (UAVs) have been extensively utilized across various domains, including product delivery, rescue operations, inspection of inaccessible areas, and security and surveillance of territory. The integration of UAVs with FL is a novel addition to autonomous systems as well as data protection. The upgraded FL allows UAVs to provide decision-making support and improve their performance. FL process has emerged as a pivotal technology in smart cities, offering a decentralized method for machine learning that ensures data privacy and security. It explores advanced technologies that empower FL, including edge computing, secure multi-party computation, homomorphic encryption, blockchain, and automated machine learning.

The cyber-physical system is one of the intelligent systems that amalgamates physical and logical devices. Physical systems can interact with the environment, and the logical system is responsible for processing and making intelligent decisions based on the physical system's interaction with the environment. It has the immense capacity to interact with the environment and process data to make intelligent systems. FL adoption in IoT holds significant promise due to its ability to control decentralized sources by preserving privacy and security. FL permits collaborative training of models beyond centralized data storage, which has several advantages over IoT applications. FL can be used by wearable health devices, including fitness trackers and smartwatches, to create customized fitness and health recommendations. These gadgets gather private medical information but they also use local model training to make inferences and forecasts like identifying irregular heart rates or anticipating patterns in physical activity.

About the Editors

S. Jayachitra is an Assistant Professor at PSNA College of Engineering and Technology, Dindigul, India. She earned her B.E. degree in electronics and communication engineering and her M.E. degree in communication systems both from Anna University, Chennai, India. She is currently pursuing her Ph.D. in information and communication engineering from Anna University. She has published more than 25 research articles in reputed international journals which are indexed in Scopus. She has won two best paper awards and has been granted 22 patents. Further, she has published more than five books with reputed publishers. With her expertise she has delivered sessions on more than 20 topics in various titles. She has 9.5 years of teaching experience, and her research interests include IoT, edge computing, healthcare, medical imaging system, 5G/6G networks.

A. Prasanth earned his B.E. and M.E. degrees in computer science and engineering (with specialization in networks) from Anna University, Chennai, India. He also earned his Ph.D. in information and communication engineering from Anna University. He has served as a recognized Ph.D. supervisor. Five scholars are pursuing their research and two completed their Ph.D.s under his guidance. His name is included in the *World's Top 2% of Scientists* list in the years 2023 and 2024, released by Elsevier and Stanford University, USA. He served as a doctoral committee (DC) member for eight Ph.D. scholars. He has published more than 45 research articles in international journals/conference proceedings among which 20 were indexed in SCI and 24 were indexed in Scopus. His *h-index* is 20 and *i10 index* is 24. Moreover, he has published 10 patents at the national and international levels. He also published 12 books with reputed international publishers. He is an Associate Professor at Vel Tech Rangarajan Dr. Sagunthala R&D Institute of Science and Technology, Chennai, India. His research interests include IoT, machine learning, wireless sensor networks, ad hoc networks, and computer networks.

Rajesh Kumar Dhanaraj is a distinguished Professor at Symbiosis International (Deemed University) in Pune, India. Before joining Symbiosis International University, he served as a Professor at the School of Computing Science and Engineering at Galgotias University, Greater Noida, India. His academic and research achievements have earned him a place among the *World's Top 2% of Scientists* list globally, a recognition bestowed upon him by Elsevier and Stanford University. He earned his B.E. degree in computer science and engineering from Anna University, Chennai, India, in 2007. Subsequently, he earned his M.Tech. degree from Anna University, Coimbatore, India, in 2010. His relentless pursuit of knowledge culminated in a Ph.D. in computer science from Anna University, Chennai, India, in 2017. He has authored and edited over 50 books on various cutting-edge technologies and holds 22 patents. Furthermore, he has contributed over 115 articles and papers to esteemed referred journals and international conferences in addition to providing chapters for several books. Dr. Dhanaraj has also delivered numerous tech talks on disruptive technologies.

He has forged partnerships with esteemed professors from top QS-ranked universities around the world, fostering a global network of academic excellence. His research interests encompass machine learning, cyber-physical systems, and wireless sensor networks. Dr. Dhanaraj's expertise in these areas has led to numerous research talks on applied AI and cyber-physical systems at various esteemed institutions. Dr. Dhanaraj has earned the distinction of being a Senior Member of the Institute of Electrical and Electronics Engineers (IEEE). He is also a member of the Computer Science Teacher Association (CSTA) and the International Association of Engineers (IAENG). Dr. Dhanaraj's commitment to academic excellence extends to his role as an associate editor and guest editor for renowned journals, including Elsevier's *Computers and Electrical Engineering*, *Human-centric Computing and Information Sciences*, Emerald's *International Journal of Pervasive Computing and Communications*, and Hindawi's *Mobile Information Systems*.

Balasubramaniam S (IEEE Senior Member) is an Assistant Professor in the School of Computer Science and Engineering, Kerala University of Digital Sciences, Innovation and Technology, Kerala, India. He has 15+ years of experience in teaching, research, and industry. He completed his post-doctoral research in the Department of Applied Data Science, Noroff University College, Kristiansand, Norway. He has a Ph.D. in computer science and engineering from Anna University, Chennai, India (2015). He has published 20+ research papers in reputed SCI/WoS/Scopus indexed journals. He was granted one Australian patent and one Indian patent and published three Indian patents. He has presented papers at conferences, contributed chapters to edited books, and has been editor of books published by international publishers. His research and publication interests include machine learning and deep learning-based disease diagnosis, cloud computing security, generative AI, and electric vehicles.

Danilo Pelusi earned a degree in physics from the University of Bologna, Italy, and a Ph.D. in computational astrophysics from the University of Teramo, Italy. Currently, he is an Associate Professor of Computer Science in the Department of Communication Sciences, University of Teramo. Having worked as the editor of books for reputed publishers, and Associate Editor of IEEE Transactions on Emerging Topics in Computational Intelligence (2017–2020), IEEE Access (2018–present), IEEE Transactions on Neural Networks and Learning Systems (2022–present), and IEEE Transactions on Intelligent Transportation Systems (2022–present), he is also the guest editor for Elsevier, Springer, MDPI, and Hindawi journals. Keynote speaker, Guest of Honor, and Chair of IEEE conferences, he is an inventor of international patents on AI. As a World's 2% Top Scientist 2021, his research interests include fuzzy logic, neural networks, information theory, machine learning and evolutionary algorithms.

Contributors

P. Arul
Sri Venkateswara College of Engineering,
India

Afizan Bin Azman
Taylors University,
Malaysia

S. Balakrishnan
Aarupadai Veedu Institute of Technology, Vinayaka Mission's Research Foundation (DU),
India

M. Balasubramani
V.S.B. Engineering College,
India

Korhan Cengiz
University of Jordan,
Jordan

Mani Deepak Choudhry
SRM Institute of Science and Technology,
India

S. Deepa
VISTAS,
India

Rajesh Kumar Dhanaraj
Symbiosis International (Deemed University),
India

Ahmed A Elngar
Beni-Suef University,
Egypt.

Indhumathi G
Rajalakshmi Engineering College,
India

Saranya G
Rajalakshmi Engineering College,
India

Syed Shahul Hameed
University of Technology and Applied Sciences – Sur Campus,
Oman

S. Jayachitra
PSNA College of Engineering and Technology,
India

Princy K T M
Amrita Vishwa Vidyapeetham,
India

Saranya K.
Dr. Mahalingam College of Engineering and Technology,
India

Anju Kulkarni
Dayananda Sagar College of Engineering,
India

S. Lavanya
Muthayammal Engineering College,
India

D. Manibharathi
Government College of Engineering,
India

A. Meenambika
PSNA College of Engineering and Technology,
India

Radhika Menon
Ajeenkya D.Y. Patil University,
India

P. Murugapandiyan
Anil Neerukonda Institute of Technology and Sciences,
India

Mariya Ouaissa
Moulay Ismail University,
Morocco

Manjula P.
King Faizal University,
Saudi Arabia

Danilo Pelusi
University of Teramo, Balzarini,
Italy

A. Prasanth
Vel Tech Rangarajan Dr. Sagunthala R&D Institute of Science and Technology,
India

Sunil Kumar RM
R.L. Jalappa Institute of Technology,
India

Raghu Ramamoorthy
The Oxford College of Engineering,
India

N. Sathish
Sri Venkateswara College of Engineering,
India

S. Simonthomas
Aarupadai Veedu Institute of Technology,
India

M. Sundarrajan
SRM Institute of Science and Technology,
India

Pham Chien Thang
TNU-University of Sciences,
Vietnam

G. Uthradevi
Chennai Institute of Technology,
India

Anitha Velu
Sri Sairam College of Engineering,
India

Mithra Venkatesan
Dr. D. Y. Patil Institute of Technology,
Pune, India

M. Yogeshwari
VISTAS,
India

V. Yokesh
Sathyabama Institute of Science and Technology,
India

1 Journey toward Federated Learning

Fundamentals, Tools Paradigms, Opportunities, and Challenges

Mithra Venkatesan, Radhika Menon, Anju Kulkarni, and S. Jayachitra

1.1 INTRODUCTION

Federated learning (FL) enables decentralized model training by allowing multiple devices or organizations to collaboratively develop a shared model without transferring their local data to a central server. FL ensures that sensitive information remains on local devices, addressing privacy concerns, while harnessing the power of distributed datasets. Instead of raw data, FL exchanges model parameters or gradients between participants, enabling collaborative learning without compromising individual or organizational data ownership. This approach aligns seamlessly with the increasing emphasis on privacy-preserving technologies in today's digital landscape [1].

The potential of FL extends beyond privacy preservation. By leveraging diverse, geographically distributed datasets, FL can improve model robustness and generalizability. For example, in healthcare, FL facilitates collaboration between hospitals to develop predictive models while safeguarding patient data. Similarly, in telecommunications, FL enables the training of models on user behavior without violating privacy norms. However, these benefits come with unique challenges, such as handling data heterogeneity, ensuring secure communication, and managing system reliability in decentralized environments.

The core principles of FL include data decentralization, privacy preservation, and collaborative intelligence. Furthermore, techniques such as differential privacy, secure aggregation, and homomorphic encryption enhance the security of shared model updates, preventing adversarial attacks and safeguarding sensitive information. This makes FL a promising framework for applications in sectors like healthcare, where institutions can collaborate on predictive models without sharing patient records, or

DOI: 10.1201/9781003532323-1

in finance, where organizations can leverage combined insights without exposing proprietary data [2].

Despite its transformative potential, FL faces significant challenges, including managing data heterogeneity, ensuring reliable communication between participants, and addressing computational constraints on edge devices. Additionally, the scalability of FL systems and their resistance to adversarial threats remain active areas of research. This chapter aims to provide a comprehensive introduction to FL, exploring its foundational concepts, tools, and frameworks, as well as the opportunities and challenges.

1.2 CONCEPT OF FEDERATED LEARNING

FL method is a knowledge strategy that is part of machine learning. This method trains the algorithm through different decentralized devices which are present either on edge or on a server having the local data and doesn't exchange the local data. This is a powerful technique that will empower multiple actors to develop a common, strong machine learning prototype in the absence of sharing of data. Various critical issues like the privacy of data, security of data, access rights of data, and mixed data access are addressed by this technique. There are humongous applications of FL covering different domains. FL also aims in using machine learning algorithms such as convolutional networks which are part of deep neural networks. However, the training is accomplished on local datasets using local networks and results in building a universal model using data jointly presented by all the nodes [3],[4].

FL thus consists of the following steps in its execution:

1. The first step is identifying the problem which is to be solved using FL and subsequently the most suitable artificial intelligence (AI) model which can be used.
2. The selection of existing online devices as participants is done arbitrarily or by a specific method in the process of learning.
3. Using the local data, each of the individual models are trained.
4. After the local training, the devices which are participating send the parameters associated with the model to a central server via a communication method.
5. The dominant server works as an aggregator by collecting various individual models and formulates the universal model.
6. The parameters of the universal model are subsequently updated in all the individual models.

1.2.1 Categories of Federated Learning

The classification is decided on majorly five distinct aspects. They are categorized based on how data is partitioned, mechanisms involved in privacy, machine learning models which will be adopted, architecture for communication, and techniques used for solving heterogeneity [4].

Based on portioning of data, the decentralized technique can be classified as horizontal, transfer, and vertical FL methods. In the horizontal method, users are kept distinct while the features associated with the users will remain common. In this method, the dataset is split horizontally where data in different rows belong to different users but have similar user features. For example, there could be two users providing similar services from different regions. Hence the users are different, but the features remain similar. In such models, horizontal FL is used [5].

On the contrary, vertical FL divides the dataset vertically, where the users remain the same and user features are split or are different. Data is present in various columns and is aligned according to the handler. The feature proportions of the training data are improved through vertical FL. To understand through an example, there could be two different users of a bank and a company involved in e-commerce. Their user group is likely to include all the people living in the locality. The bank records users' income, expenditure and the company has details associated with user's record purchasing and browsing history. Hence, though users are common, there is no intersection of users' features. Vertical FL will bring together and integrate all these features to enhance the learning ability of the prototype developed [6].

Federated transfer learning is popularly used in situations where the users, as well as user features, are distinct and have no scope to remain in common. For example, we can consider an industry involved in e-commerce in one country and a community application in another country. Because of users from different geographical locations, the user groups will not be common. Also, because of variation in the type of users the attributes of the users will also be distinctly varied. Effective FL in such types of models is carried out with the help of federated transfer learning. Problems are solved using this method when there is very little data size as well as small label samples. The most apt situation for the usage of this type of method is when optimization of the performance of a task is required but sufficient or related data is not available. This technique could be employed in radiology departments to build radiology diagnosis systems using different x-ray scans. Through this method privacy-involving data is protected as well as supplementary tasks are transferred to the learning director and help us unravel the problem involving minor amounts of data [7].

In the application setting of FL, there are different difficulties such as an uneven data distribution, the computing power of equipment etc. Further, when there is communication involved, the model should be chosen such that there is no leakage of sensitive information. There are different FL models proposed to improve the communication architecture such as federated averaging, federated dropouts working toward faster convergence and communication between the server and the device.

In applications during usage of FL, it is observed that the different usage of equipment or devices will affect the competency of the complete process involved in training. These would be the problems arising due to heterogeneity. Toward the same, different solutions are proposed such as communication asynchronously, sampling of the device, a mechanism for fault tolerance, and heterogeneity model. Since the devices are scattered in FL, asynchronous communication will be the scheme used [8].

FL has also been classified based on aggregation algorithms that are being used. There are different approaches such as federated average (FedAvg), secure aggregation, federated proximal, and federated meta-learning algorithm (MA). The simplest approach is the federated average which was developed by Google. In this approach, the coordination is performed centrally by the server and the universal parameters are revealed to some clients. The local models will train their inputs and are given to the server. The global prototype will average updates which it gets from the local models. Secure aggregation focuses more on aggregation along with providing security and privacy. This is especially useful for a mobile environment that is participant based. An improvised version of federated average taking into account non-uniformity and heterogeneity is the next approach called federated proximal. Federated MA is popular for deep neural networks and models based on short-term memory [9].

1.2.2 Degrees of Federated Learning

Degree of FL is classified based on the number and type associated with participating devices. The two types based on the degree are cross-silo and cross-device. When there are a small number of participating entities, the type of FL is labelled as cross-silo. In cross-silo, there are two to hundred entities or devices belonging to different clients comprising hospitals, institutions, schools etc. Low cost in spite of the heterogeneity involved is characterized in the cross-silo model. Cross-device model is at a much larger scale as well as has greater heterogeneity as compared with cross-silo. The number of heterogeneous devices participating can range in millions. There is a need of huge computational power to handle these models. Further, these models are also vulnerable to attacks as well as privacy issues [10].

1.2.3 Topologies in Federated Learning

The structure in which individual entities or devices are connected is very important in FL. The broad categorization of topology is centralized FL and decentralized peer-to-peer FL. In centralized architecture, a single dominant server takes up the role of the collector comprising the universal model, which collects all the trained local models. However, in a decentralized peer-to-peer model, there is no central server. The machines communicate among each other. Leader election results in identifying a leader node and participating or follower nodes. The potential risks associated with central server such as operational failure and susceptibility to attacks are absent in this model. These decentralized models could further have ring topology or hybrid topology. In hybrid topology, the participating nodes are grouped and arranged in accordance to the use cases. In contrast, ring topology has participating nodes arranged sequentially. Centralized models have better scalability as compared with decentralized models. However, tolerance to faults as privacy are greater concerns in decentralized approaches in comparison with centralized approaches. Communication overhead is also reduced in centralized models [11].

1.2.4 Standardization Efforts

IEEE 3652.1-2020 is the guide of IEEE which has been developed recently for the architectural outline and FL-based application. This provides a blueprint on data that could be used and models that could be built across various devices as well as organizations adhering to the constraints on security and privacy and regulatory requirements as mentioned in this standard. It details the different architectural frameworks along with guidelines for applications for federated machine learning. This includes definitions, descriptions, and classification of federation learning along with scenarios in which it can be applied in different domains. The performance evaluation of these learning algorithms is also detailed with the requirements along with performance metrics [12].

1.3 RECENT DEVELOPMENTS IN FEDERATED LEARNING

The recent developments in FL are toward building models capable of personalization. Personalization involves understanding the individual needs of the users and building models to suit their needs and offer the best services accordingly. Personalization becomes relevant for varied domains starting from development of personal assistant to making advanced gaming consoles. Accomplishing personalization takes time and hence current developments in FL are on building models which take less time and give a good personalized experience. Such an emerging model is federated reinforcement learning architecture.

Further, separate models exclusively working toward attack detection and protection of privacy are also fast emerging involving blockchain. The usage of blockchain for preventing intrusion risk in a central server is finding applications in varied areas such as in smart city infrastructure, vehicular cyber–physical systems, and 5G heterogeneous networks. Potential vulnerability assessment is done and incorporated into the models of FL. Further, FL can be adapted and integrated with existing learning frameworks and algorithms involving supervised, semi-supervised, unsupervised, and reinforcement learning settings. Such intersections of FL models with existing learning frameworks are gaining usage, which is termed as federated learning [13].

Federated transfer learning is also put under the same umbrella with a technique called knowledge distillation where the servers present centrally do not control the localized models. Separate models are trained in every client end along with some structure shared among the models. However, learning across local datasets at various clients is considered an individual task. This emerging technology is federated multi-task learning [14].

Also, federated meta-learning is a strategy widely evolving where the prototype is capable of adapting quickly to new tasks with limited training data available when the client is capable of handling different types of learning tasks. FL techniques are also combined with generative adversarial networks (GANs) where the consumers are trained in the vicinity, and the fusion of the global model is done with various strategies. GANs generally consist of two competing models which consist of a generator

TABLE 1.1
Filtering in Scopus Search Engine

Filtering Criteria	Reject	Accept
Search criteria		
Search engine: Scopus		
Search date: 4 August 2024		
Search term: "Federated Learning"	–	16346
Year: 2014–2024	186	16160
Subject area: Engineering	8954	7206
Document type: "Article"	3774	3432
Language: English	119	3313
Erroneous record refinement:	05	3311

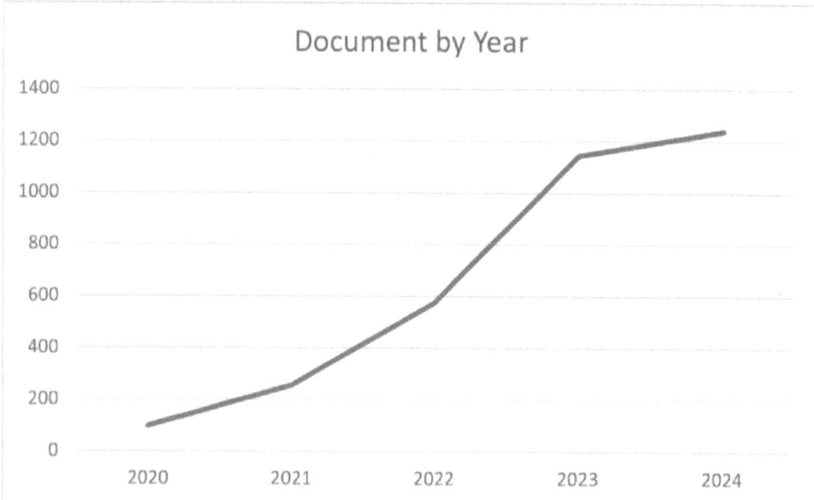

FIGURE 1.1 Number of papers over the years on federated learning.

and discriminator. The generator is capable of constructing samples and the discriminator is generally a binary classifier [15].

Throughout the globe, there is considerable interest and work which has been done on FL. To find the volume of research in the area of FL, a search and analysis was done in Scopus database on FL for the period from 2014 to 2024. Through the varied filter process, a total of 3311 articles were filtered. The number of articles in the Scopus database along with the number of articles accepted as well rejected is detailed in Table 1.1. The quantity of papers each year on the theme of FL is shown in Figure 1.1. As it can be observed from the figure, the maximum amount of work on the theme has been done in 2024 [16–18].

The research on FL has been done by varied authors, and the number of papers by different authors on topics under study is depicted in Figure 1.2. The top 15 authors in the domain of FL are illustrated. Niyato D. has a maximum number of 37 documents to his credit.

The number of documents from different sources based on FL also varies every year. The different sources or the journals along with the number of documents per year is detailed in Figure 1.3.

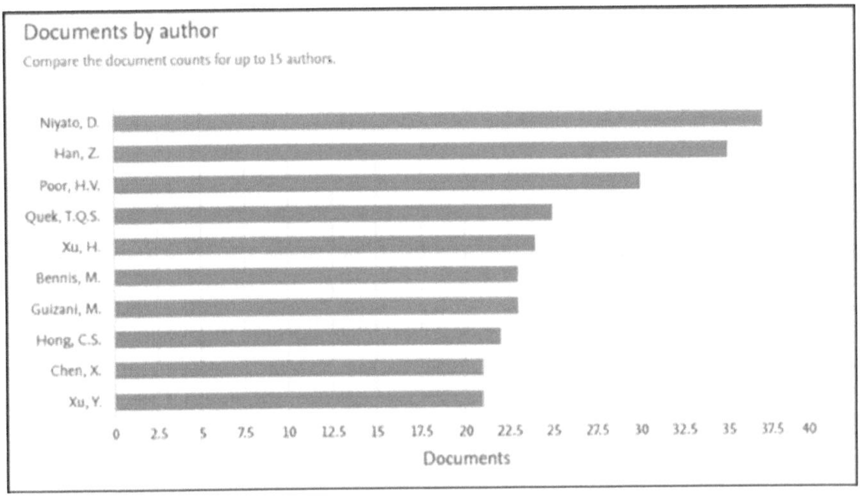

FIGURE 1.2 Number of papers by different authors on federated learning.

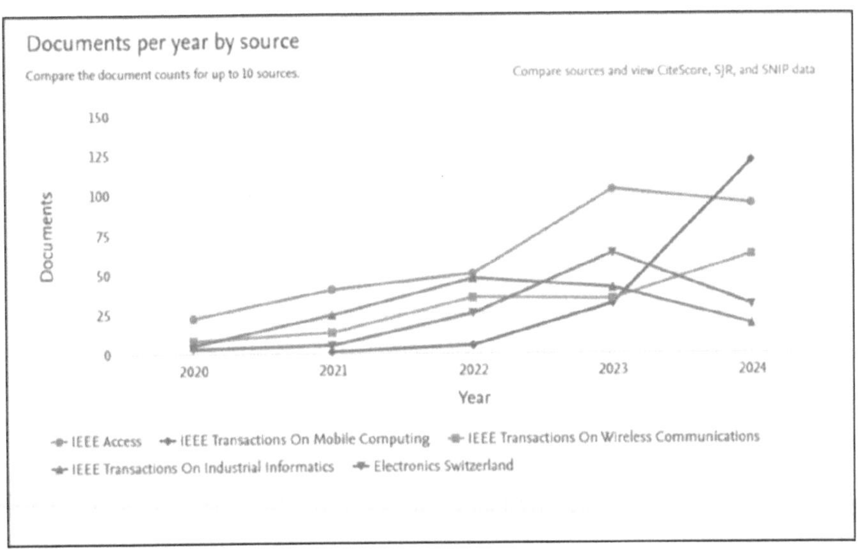

FIGURE 1.3 Documents per year by source.

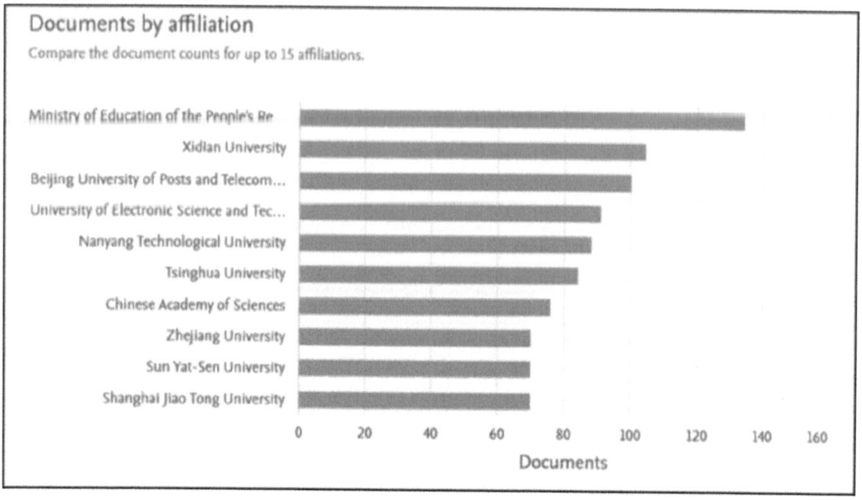

FIGURE 1.4 Number of documents per year by affiliation.

The analysis was also done on the number of documents available in accordance to the affiliation. This gives an idea on the various research groups and universities working on the topic around the globe. This is shown in Figure 1.4. It is found that the Ministry of Education of the People's Republic of China has maximum number of 134 documents by affiliation, while the least is Shanghai Jiao Tong University with 70 documents [19, 20].

The analysis on documents by country on the topic of FL is also done, with China having a maximum of 1653 documents followed by the United States with 517 documents [21]. This analysis on documents by country is illustrated in Figure 1.5.

From all the graphs it is commonly observed that this topic is gaining huge relevance and popularity in recent years and can be utilized for a variety of applications. The country-wise network density diagram on FL is given in Figure 1.6. This gives a comprehensive idea on how certain countries like United States and China are dominating the research on FL [22].

The organization-based network density diagram on FL is depicted in Figure 1.7. This gives an overview of different organizations around the globe involved in research on FL [23].

There are varied authors who dominate research on FL as demonstrated in Figure 1.8. This helps not only identify who is working in the area but will also enable collaborative research in the domain of study. Figure 1.9 details the network density diagram on FL based on keywords. The relation and linking of the keyword "federated learning" with other words is also shown in Figure 1.10.

Many papers are cited together in the source articles. The mechanism which tries to find such papers that are cited together is called co-citation analysis. When the same sets of papers are cited by many authors, it results in formation of clusters. Among these clusters or groups formed, there could be common themes that can

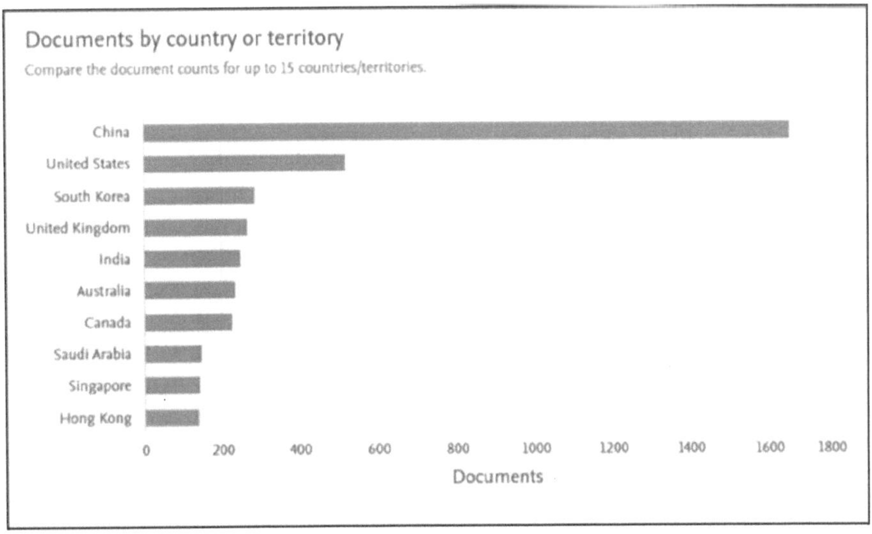

FIGURE 1.5 Number of documents by country.

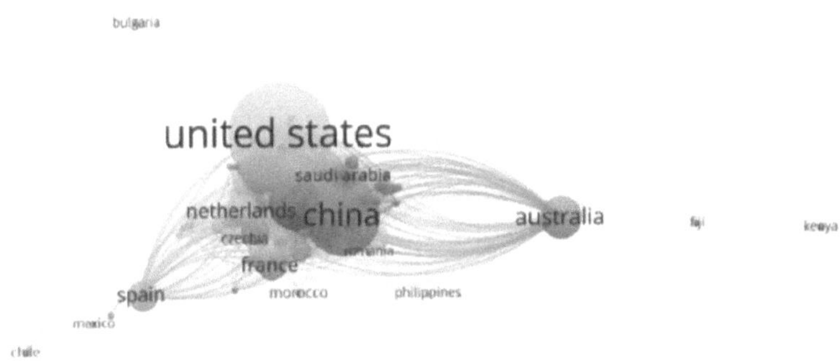

FIGURE 1.6 Country-wise network density diagram on federated learning.

emerge. These common themes could lead to potential areas where further research could be carried out.

The main intent through this analysis is to find specific areas in FL where research is being carried on and also find the areas where there is scope for research to be done. All these diagrams depict the reach of FL and hence the scope of work available in the area under study [24].

1.4 TOOLS AND FRAMEWORK USED IN FEDERATED LEARNING

FL is a distributed learning paradigm which maintains localized data while facilitating training of the model over different dispersed entities or computers [25]. This

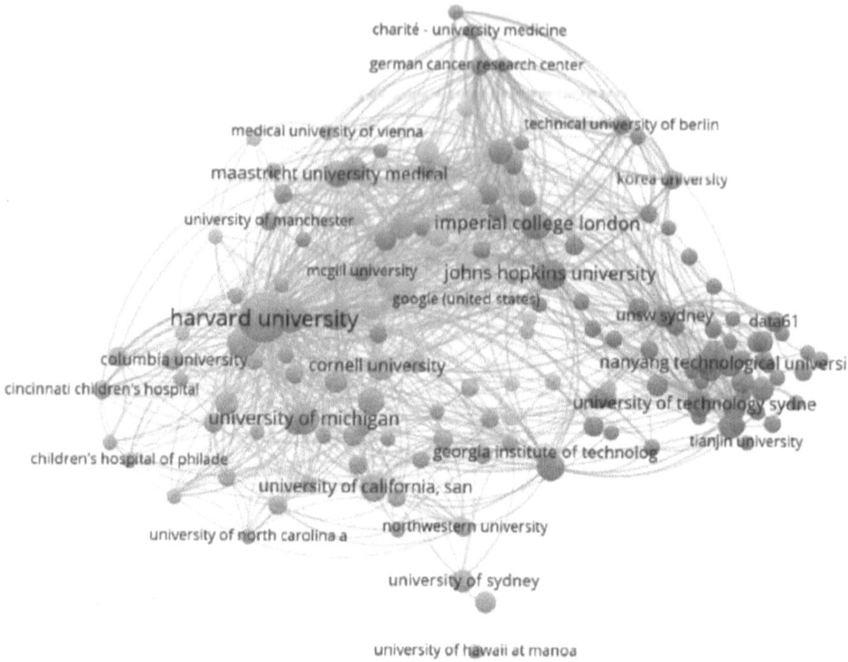

FIGURE 1.7 Organization-based network density diagram on federated learning.

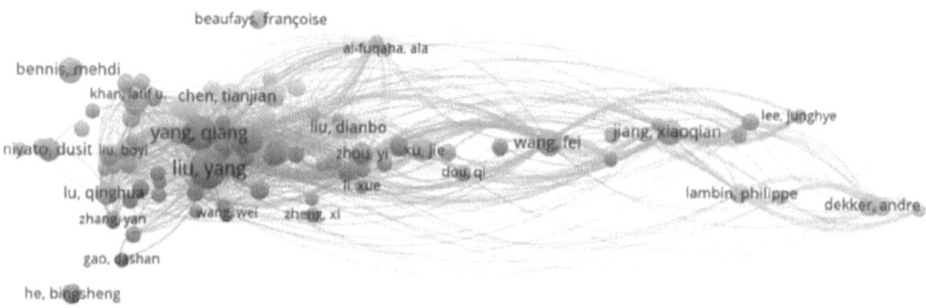

FIGURE 1.8 Author-based network density diagram on federated learning.

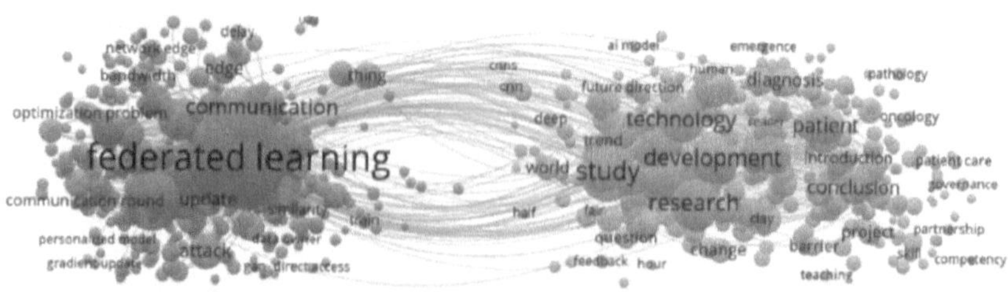

FIGURE 1.9 Keyword-based network density diagram on federated learning.

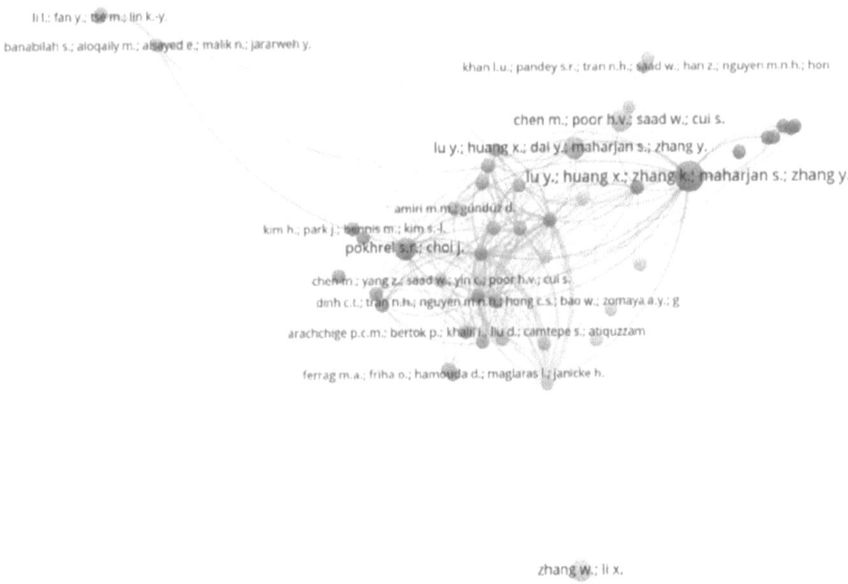

FIGURE 1.10 Co-citation analysis.

method improves efficiency and privacy, especially when sensitive data is dispersed across multiple locations. There are a number of frameworks and tools that are available for usage [26].

Federated TensorFlow (TFF) Overview

Google created the open-source TensorFlow Federated framework, which expands TensorFlow to provide FL. It offers high-level abstractions for data processing and federated computations. The application program interfaces (APIs) present enable FL algorithms to be designed and implemented. It also provides scenarios for simulation offline. There is also provision for the use of tensor flow models in a federated environment by seamlessly integrating with tensor flow.

PySyft

PySyft is an open-source package that provides privacy-preserving machine learning and FL capabilities to PyTorch and TensorFlow. Differential privacy and safe multiparty computation capabilities are among the privacy features. It is compatible with multiple machine learning frameworks and backends, it supports FL. The adapters are available to enable integration with TensorFlow, while PyTorch integration is the default configuration.

Federated Learning by IBM

With an emphasis on enterprise use cases, IBM provides a FL solution as part of its range of AI and data science capabilities. It is made to work with both IBM's on-premises and cloud solutions. The platform places a strong emphasis on features designed specifically for enterprise data privacy and security. Further, it also facilitates extensive deployments of FL.

Microsoft Seal

Although homomorphic encryption is not a feature of Microsoft SEAL (Simple Encrypted Arithmetic Library), it can be employed in FL settings. It allows for secure FL in situations where privacy is essential by supporting encrypted calculations. Further, to improve data security, it can be paired with FL frameworks.

The Horizon

Horizon is a platform for large-scale reinforcement learning that was created by Facebook (now Meta). It can also be modified for FL. It is suitable for extensive machine learning settings. The platform is adaptable to several learning contexts, including FL environments.

Launch FL

Open Federated Learning, or Open FL, is an open-source platform made to make FL installations and experiments easier. It works with several machine learning frameworks and is framework agnostic. It permits FL workflows and techniques to be customized.

KubeFlow

Supporting FL through several components and integrations, KubeFlow is a machine learning toolkit for Kubernetes. Kubernetes Integration uses Kubernetes to handle workloads related to FL. It supports a variety of machine learning technologies and frameworks, indicating extensibility. A variety of features and integrations are offered by these frameworks and solutions to facilitate FL in various contexts and use cases. A tool's scalability, ease of integration with current systems, privacy and security requirements, and underlying machine learning framework should all be taken into account.

Substra

This is an enterprise solution exclusively made to cater to healthcare applications. It has a variety of interfaces along with python library, graphical user interface, and it can be used by a wide spectrum of users. Through encryption strategies, the framework maintains privacy and security and hence is most suited for the healthcare domain. This framework is made of deep learning models.

PyTorch

This is a popularly used machine learning framework developed by the AI group in Facebook. Through the usage of python, formulation and training of neural network models is accomplished. This framework is flexible and can support deployment on a large scale on mobile devices as well for different types of data partitions. Security and privacy are important attributes incorporated in this framework.

NVIDIA Clara

This platform empowers developers to build real-time solutions on GPU-based frameworks which are secure. This platform is popular in healthcare scenarios and makes use of auto-ML and DL-based models incorporating privacy.

1.5 BENEFITS AND LIMITATIONS OF FEDERATED LEARNING

The major advantages of using FL over conventional machine learning techniques are listed further.

1. Efficient use of hardware: Since FL makes use of data from different servers, there is no requirement of a centralized server and hence reduces the need for the usage of complex hardware.
2. Improved data security: All training data are kept in respective devices, hence improved security of data is achieved.
3. Diversity of data: Heterogeneous data can be accessed from varied devices and sources and can be processed.
4. Learning continually: As there is no aggregation of data, the model is real time and will involve continual learning.
5. Privacy of data: This is the major advantage of FL. This is mainly because various privacy enhancement strategies such as differential privacy, homographic encryption are utilized to enhance privacy of data. However, since aggregation of data is not on a single server and involves many devices for aggregating data as well as for analysis, there is an increased attack surface in this strategy.
6. Enhanced scalability: Since different devices are learning from each other, the entire network becomes scalable.
7. Reduced consumption of power and latency: FL involves creating local models, and hence compared with traditional central networks power consumption and latency is reduced considerably.
8. Improved time of training, accuracy, and cost: Training happens across local models and then are aggregated and hence training time and cost involved is also reduced. This improves the accuracy of the federated models.
9. Minimization of data: The local models sent out are temporary and get merged in the central model. Subsequently, the local model is discarded and hence the raw data present in local networks remain hidden. This results in the development of a competent and strong complete system.

Though the advantages of FL surpass the limitations, understanding the limitations is essential before detailing the usage and implementations. Some factors of the performance limitations of FL are as follows:

1. Heterogeneity of data: Though heterogeneous data can be aggregated with FL models, the heterogeneity of data may prevent or act as a deterrent to the development of generalized models and may also bring down the accuracy of the prototype developed.
2. Centralization: Though federated models are decentralized, there are still some elements of centralization where the central model makes use of output from various devices to build a new model.

3. Indirect leakage of information: Though data privacy and security is improved in this model, the model is also prone to attacks through hidden backdoors and can result in indirect leakage of information from the source.
4. Insufficient amount of data: Since the model is training locally, sometimes the quantity of data required to train becomes insufficient.

1.6 CONCLUSION

The development of FL has advantages over different applications with machine learning, deep learning, and AI. The FL uses collaborative learning and provides data privacy to the users. FL aims to build a global optimal model, taking the data from different participants that satisfies the local objective. When the data remains localized, the FL occurs with improved privacy and regulatory compliance while nurturing collaboration across diversified IoT devices. Over the deployment of FL in practical applications, some constraints arise during the training process. This can affect the efficiency and competency of the model developed. Also, FL is susceptible to attacks which reduces the accuracy and efficiency of the developed model. The ongoing research and development in FL will undoubtedly shape the landscape of AI, paving the way for innovations that prioritize both performance and privacy.

REFERENCES

1. Zhang, C., Xie, Y., Bai, H., Yu, B., Li, W. and Gao, Y., A survey on federated learning. *Knowledge-Based Systems*, *216*, 106775:1–19, 2021.
2. Yang, Q., Liu, Y., Chen, T. and Tong, Y., Federated machine learning: Concept and applications. *ACM Transactions on Intelligent Systems and Technology (TIST)*, *10*(2), 1–19, 2019.
3. Jatain, D., Singh, V. and Dahiya, N., A contemplative perspective on federated machine learning: Taxonomy, threats & vulnerability assessment and challenges. *Journal of King Saud University-Computer and Information Sciences*, *34*(9), 6681–6698, 2022.
4. Hu, K., Li, Y., Xia, M., Wu, J., Lu, M., Zhang, S. and Weng, L., Federated learning: a distributed shared machine learning method. *Complexity*, *2021*(1), 1061663:1–15, 2021.
5. He, C., Li, S., So, J., Zeng, X., Zhang, M., Wang, H., Wang, X., Vepakomma, P., Singh, A., Qiu, H. and Zhu, X., Fedml: A research library and benchmark for federated machine learning. *arXiv preprint arXiv:2007.13518,* 2020.
6. Sekar, J. and Aruchamy, P. A novel approach for heart disease prediction using hybridized AITH2O algorithm and SANFIS classifier. *Network: Computation in Neural Systems*, *36*, 1–39, 2024.
7. Lathika, V., Arunkumar, S., Balasubramani, M. and Kalaichelvi, V. Secure multi-party computation for collaborative data analysis in network security. *International Conference on Intelligent and Innovative Technologies in Computing, Electrical and Electronics (IITCEE)*, 1–5, 2024. doi 10.1109/IITCEE59897.2024.10467913.
8. Hei, X., Yin, X., Wang, Y., Ren, J. and Zhu, L., A trusted feature aggregator federated learning for distributed malicious attack detection. *Computers & Security*, *99*, 102033:1–18, 2020.

9. Kairouz, P., McMahan, H.B., Avent, B., Bellet, A., Bennis, M., Bhagoji, A.N., Bonawitz, K., Charles, Z., Cormode, G., Cummings, R. and D'Oliveira, R.G., Advances and open problems in federated learning. *Foundations and Trends in Machine Learning, 14*, 1–210, 2021.
10. Nigam, C., Aeron, A. and Suganthi, D., Real-time object detection on edge devices using mobile neural networks. *International Conference on Intelligent and Innovative Technologies in Computing, Electrical and Electronics (IITCEE)*, 13, 1–4, 2024.
11. Sudha, M. and Reddy, V. M., Optimizing intrusion detection systems using parallel metric learning, *Computers and Electrical Engineering*, 110, 108869: 1–15, 2023.
12. Paragliola, G. and Coronato, A., Definition of a novel federated learning approach to reduce communication costs. *Expert Systems with Applications, 189*, 116109:1–15, 2022.
13. Siddiqa, A., Khan, W. Z., Alkinani, M. H., Aldhahri, E. A. and Khan, M. K., Edge-assisted federated learning framework for smart crowd management. *Internet of Things*, 101253:1–19, 2024.
14. Veerasamy, V., Qiu, H., Ghias, A. M. Y. M., Chauhan, K., Nguyen, H. D. and Gooi, H. B., Federated-learning-based distributed frequency control against false data injection attack. *IEEE Transactions on Industrial Electronics, 71*: 8981–8992, 2023.
15. Song, X., Chen, Z., Wang, J., Zhang, Y. and Sun, X., A multi-stage LSTM federated forecasting method for multi-loads under multi-time scales. *Expert Systems with Applications, 253*, 124303:1–16, 2024.
16. Huang, Y., Huo, Z. and Fan, Y., DRA: A data reconstruction attack on vertical federated k-means clustering. *Expert Systems with Applications, 250*, 123807:1–21, 2024.
17. Lakhan, A., Mohammed, M. A., Abd Ghani, M. K., Abdulkareem, K. H., Marhoon, H. A., Nedoma, J., Martinek, R. and Deveci, M., FDCNN-AS: Federated deep convolutional neural network alzheimer detection schemes for different age groups. *Information Sciences*, 120833:1–18, 2024.
18. Zhang, J., Zhu, H., Wang, F., Zheng, Y., Liu, Z. and Li, H., iDP-FL: A fine-grained and privacy-aware federated learning framework for deep neural networks. *Information Sciences*, 121035:1–14, 2024.
19. Li, J., Wei, H., Liu, J. and Liu, W., FSLEdge: An energy-aware edge intelligence framework based on federated split learning for industrial internet of things. *Expert Systems with Applications, 255*, 124564:1–14, 2024.
20. Wang, R., Huang, W., Lu, Y., Wang, J., Ding, C., Liao, Y. and Shi, J., Cloud-edge collaborative transfer fault diagnosis of rotating machinery via federated fine-tuning and target self-adaptation. *Expert Systems with Applications, 250*, 123859:1–13, 2024.
21. Xu, H., Fan, Z. and Liu, X., Application of personalized federated learning methods to environmental sound classification: A comparative study. *Engineering Applications of Artificial Intelligence, 135*, 108760:1–18, 2024.
22. Chen, H., Hsu, H. Y., Hsieh, J. Y. and Hung, H. E., A differential privacy-preserving federated learning scheme with predictive maintenance of wind turbines based on deep learning for feature compression and anomaly detection with state assessment. *Journal of Mechanical Science and Technology, 38*(7): 3413–3429, 2024.
23. Kumar, G. and Toshniwal, D., FedNISP: Neuron importance scope propagation pruning for communication efficient federated learning. *Computers and Electrical Engineering, 118*, 109349:1–19, 2024.
24. Begum, K., Mozumder, M. A. I., Joo, M. I. and Kim, H. C., BFLIDS: Blockchain-driven federated learning for intrusion detection in IoMT networks. *Sensors, 24*(14): 4591–4604, 2024.

25. Liu, F., Jiang, A. and Chen, L., A multi-scale channel attention network with federated learning for magnetic resonance image super-resolution. *Multimedia Systems*, *30*(4): 214–232, 2024.
26. Tang, X., Lan, T., Zhong, H., Li, D. and Miao, Q., An evolutionary game model for indirect data sharing in manufacturing big data consortium. *Expert Systems with Applications*, *255*, 124807:1–18, 2024.

2 Federated Learning-Based Algorithms for Deployment and Model Optimization

S. Jayachitra, A. Meenambika, M. Balasubramani, and Korhan Cengiz

2.1 INTRODUCTION

Federated learning (FL) is a decentralized machine learning (ML) algorithm which can function either entirely in decentralized or centralized platforms. With FL, a model can be concurrently trained across several devices or nodes at the edge without gathering raw data in a single location [1]. This technique was initially introduced by Google that uses the data of millions of users to train a common machine learning (ML) technique while maintaining their privacy [2]. The application of models on the edge and cooperative training on distributed datasets offer several significant benefits. The raw data does not have to be removed from the device which improves security of the data. Resources are free since a huge centralized server with strong computing power is not required, and distributed data processing storage drastically lowers bandwidth and energy consumption. Real-time continuous learning is encouraged concurrently because interpretation can be done without requiring communication with a centralized server as models are located on the edge. These features enable artificial intelligence (AI) systems to be scaled effectively through deploying dispersed nature of the internet of things (IoT) [3].

Several businesses have adopted FL integration in IoT through providing innovative, secure, and intelligent solutions across multiple domains [4]. It involves intelligent and self-adapting vehicles that strengthen performance, safety, and production that can adjust to altering conditions naturally. The healthcare AI applications can protect secret patient data, customized financial services, fraud detection, and traffic management for feasible urban environments [5].

The amalgamation of FL, AI, and IoT are pushing towards new territories of analysis and processing. These inventions cause a paradigm change in technology through redefining the relationship between humans and machines thus transforming the industry [6]. Each of FL's various application subjects and groups has unique problems that call for particular methods to solve them. The existing approaches offer

DOI: 10.1201/9781003532323-2

a range of techniques for dealing with these issues. The categories, taxonomy, application domains, model optimizations are delineated in this chapter.

2.1.1 Key Technologies

FL offers a combination of ML techniques that provides a disseminated collection of sources of information to collectively instruct a shared model. Alternatively, personalization in FL relates to the recursive improvement of an aggregate ML technique for the intention of formulating to the particular features and prerequisites of source data. The steps include altering the parameters of the model and hyperparameter approaches to improve the accuracy of technique on user-specific information. Alternatively, personalized federated learning (PFL) defines an instance of FL focused on developing ML models that are optimized for specific features and information of each client in a federated network. The aim of PFL is to create ML models that can be modified and offer customized predictions or recommendations based upon each client's local information while maintaining data confidentiality and privacy [7].

2.2 DEPLOYMENT STRATEGIES FOR FEDERATED LEARNING

The deployment strategy of FL can be classified into two taxonomies: cross-device FL (C_DFL) and cross-silo FL (C_SFL) techniques, which are exploited for numerous applications, namely, smart healthcare, smart cities, smart vehicles, and unmanned aerial vehicles (UAVs).

2.2.1 Cross-Device Federated Learning (C_DFL)

C_DFL is a distributed ML algorithm where it is trained over numerous devices while maintaining the data bounded on every device. It is widely deployed in various environments where data security is crucial, namely, IoT, smartphones, and military applications. The C_DFL can be used for environments necessitating various learning mobile devices. Particularly, the learning exists over multiple devices using data obtained through specific users. It can be utilized on wearable applications such as heart-rate monitoring, glucose-level monitoring. It is also employed in mobile applications such as Google's Gboard where the system is trained on specific user data to enhance typing experience. Here, the IoT device plays a key role in enhancing the model's performance by managing efficiency and traffic. The different resources and diversified dissemination of data are the major challenges in C_DFL parameter settings. In other words, the primary challenges in C_DFL involve heterogeneity of data and communication efficiency [8].

2.2.2 Cross-Silo Federated Learning (CsFL)

A decentralized method of ML called CsFL involves several organizations, or "silos," working together to jointly train a common ML model without exchanging data.

With this approach, the organization may take advantage of aggregate data without sacrificing data privacy. There is a notable difference between the architecture and training framework of CsFL. Under this scenario, user devices may exchange model parameters as well as particular intermediate results, depending on the specification of the training procedure. This cooperative sharing makes gradient computations easier for other users, maybe using a central server that serves as an impartial intermediary. Either horizontal federated learning (HFL) or vertical federated learning (VFL) can be utilized to store the training data. CsFL is widely used, especially in circumstances where several big firms are involved.

Instead of treating small devices, the C_SFL deals with silo data, which is often found in organization firms like banks and hospitals. Since rules frequently forbid the release of this data, C_SFL seeks to protect privacy while training shared models. The contributing clients are stable, reliable, and limited which is contrary to cross-device settings. Nevertheless, the equal dissemination of training samples may not be guaranteed [9]. The performance of C_DFL and CsFL is delineated in Table 2.1.

TABLE 2.1
Comparative Analysis of C_SFL and C_DFL

Parameters	C_SFL	C_DFL
Participant scope	Focuses on a small number of important organizations, including businesses or universities	Includes a range of personal gadgets, such as wearable technology and cell phones
Data contribution	A significant amount of data is contributed by every organization	Each device provides a limited amount of information
Computational capabilities	Organizations possess substantial computational resources	Devices have a limited amount of processing power
Privacy and security measures	Focuses on protecting data confidentiality	Increased privacy threats as a result of data inversion, model extraction, and membership inference
Collaboration requirements	Calls for encouraging teamwork and offers rewards for involvement in the organization	Requires the development of incentive programs and the guarantee of fair participation
Use case scenarios	All participating entities have the same data structures	Ideal for situations with limited computational resources and a large number of dispersed data sources
Challenges	Focused on user diversity, privacy, data security constraints, and incentives for collaboration	Must handle issues with computing constraints, privacy violations, and maintaining incentives and fairness

2.3 APPLICATIONS OF C_DFL AND C_SFL

FL has been exploited in various applications across several industries. It functions by storing data locally and harmonizing the model's training process across several devices using the advantage of edge computing. The merits comprise data security, privacy, and communication efficiency in real-world applications. Subsequently, these merits promote the adoption of FL technology that manages sensitive information across a range of sectors. Numerous applications of FL and its effectiveness and strengths are given as follows:

2.3.1 Healthcare

In healthcare applications, data privacy plays a crucial role in hospitals. By combining training models without sharing the clients' medical information, FL has been applied to customized health guidance, prediction of disease, and health-monitoring systems [10]. It can identify the problems where the taxonomy has a direct impact and consequently the solutions needed by examining the features of two applications in the healthcare sector. The categories related to each application and the issues that affected it are shown. A cross-device with non-synchronous, horizontal, and distributed application is shown in Figure 2.1. The cross-silo device including horizontal, synchronous, and localized environments is illustrated in Figure 2.2.

2.3.2 Smart Cities

New constraints relating to traffic, health, natural environment, energy, and many other areas have emerged as a result of the metropolitan population's explosive growth. Many applications have surfaced in conjunction with IoT to tackle these issues by uncovering hidden patterns and thriving insightful information from IoT data. FL is preferred for this domain since sensors in smart cities gather data from the public which is essential to adherence to the General Data Protection Regulation. Further, the communication overhead can be decreased through not gathering the enormous

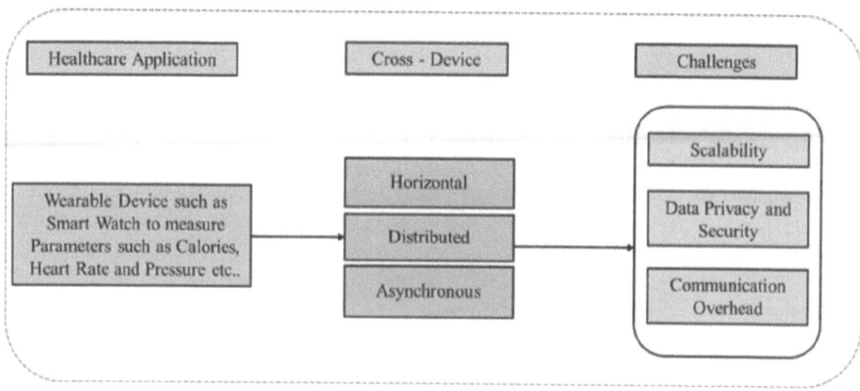

FIGURE 2.1 Cross-device healthcare applications and challenges.

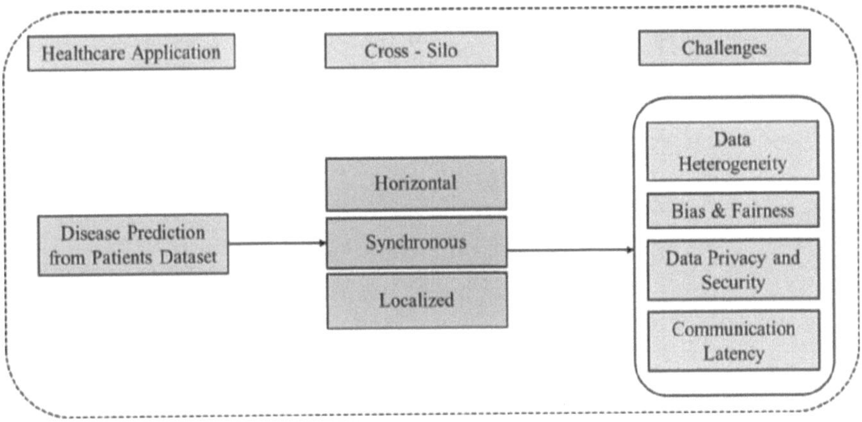

FIGURE 2.2 Cross-silo healthcare applications and challenges.

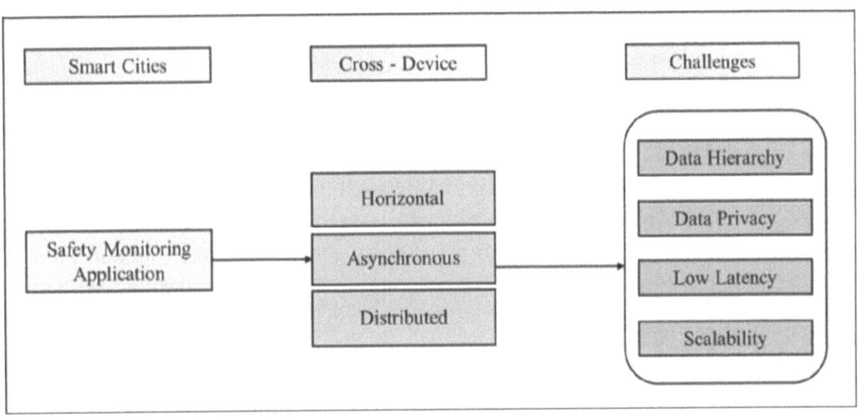

FIGURE 2.3 Smart cities safety monitoring applications.

volume of data which is generated by sensors. It also improves network capacity in addition to security. A computer-vision-based FL system for safety monitoring is described in Reference [11]. The taxonomy and associated difficulties for this use case is illustrated in Figure 2.3.

2.3.3 Smart Vehicles and Unmanned Aerial Vehicles

Analogous to earlier areas, smart vehicles collect private information that needs to be kept private, like location and driving habits. Similarly, vehicles must be able to adjust in real time to changes in road conditions. Moreover, smart vehicles can use collaborative intelligence to enhance their models with unknown knowledge when there are numerous smart vehicles on the road and sensors in smart cities. It is clear that FL can be used to increase the efficiency of smart cars. Applications in this area

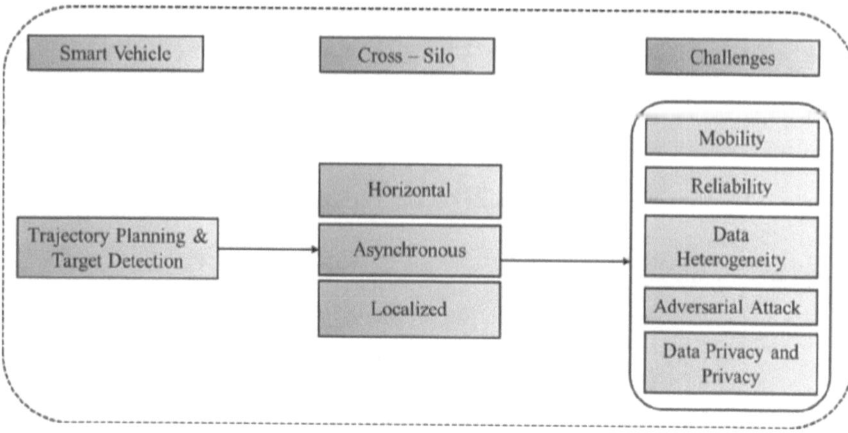

FIGURE 2.4 Target recognition in unmanned aerial vehicle.

include energy management of electric vehicles, driver behavior evaluation, vehicle trajectory, and collision prediction.

A number of traits of unmanned aerial vehicles (UAVs), quickly expanding subdomain of smart vehicles, have encouraged the use of FL in this field. Since UAVs frequently collect sensitive data, privacy is a major constraint. Hence, the new techniques should be able to safeguard them. Due to limited transmission bandwidth, the UAVs do not have the ability of transferring large volumes of data to a centralized server. Since UAVs typically operate in a variety of applications, their models must be flexible enough to adjust to change in circumstances. UAVs can coordinate with other UAVs to take part in co-operative missions.

FL provides an appropriate solution where privacy is protected by design, raw information stays on the device and cost of the communication is decreased. The repetitive learning helps the UAVs to adjust to environmental changes and jointly enhance their information usage from other UAV data. The numerous uses are cooperative image categorization, coordinated trajectory planning, shared planning, and handling of resources. The taxonomy and issues must be resolved in a trajectory planning and collaborative target identification application which are discussed in References [12,13] and demonstrated in Figure 2.4.

2.4 CATEGORIZATION AND TAXONOMY

FL has been classified based on the data distribution or the entities. Though the current research has presented a new algorithm which widens this classification to include the aggregation pattern of FL.

2.4.1 Data Distribution Classification

The data distribution of FL depends upon the segregation and dissemination nature of data over participating nodes. The datasets consist of samples and their associative attributes. Attributes have the measurable property and features of the instance, specifically related to the columns of a dataset. The instances have data points and are relative to the rows of the dataset.

2.4.1.1 Horizontal Federated Learning (HFL)

In horizontal federated learning (HFL), the correlation persists in the data attributes which proliferate through different nodes. However, that information is divergent in sample space. The existing techniques are focused on numerous applications of smart IoT devices. The data from the sample space may vary but have correlated attributes. In addition, the diverse HFL was developed to meet the requirement for the confined entities. Transforming the diverse areas can resolve the constraints of confined shortage adopted numerous times utilizing every participant as the intent domain [14].

In HFL, the nodes disseminate the same attribute space but do not utilize the same instance space. For instance, assume two hospitals in different areas play a key role in improving the disease detection model, thus trying to improve a disease prediction model. For this case, both hospitals have data such as age, gender, symptoms, and test results from the dataset but every hospital handles different patients via different instance space. Practically, the HFL involves engaging numerous devices rather than following the same dissemination in each device.

2.4.1.2 Vertical Federated Learning (VFL)

The vertical federated learning (VFL) is applicable for the data partitioning in upright direction according to the attribute dimension. The information can be categorized for multiple sets where the sets have monotonous data. The data can be coincided partially on instance ID but they are variant from attribute space. For instance, the hospital aims to detect diabetes. Many researchers found that the people with higher blood pressure and rotundity are probable to have type-2 diabetes, through analyzing the age and weight of the patient by providing imprecise dimensions eternally with past history of medical records. Although the younger-aged patient who intakes more sugar as well as is lacking physical activity but not much obese also suffers from diabetes. This prediction is based on personalized owing to the lack of information. The advanced technology such as smartphone application as step counters and dietary planning may correlate with each other. The VFL nodes distribute the same instance space but not the same attribute space [15].

As another instance, assume the manufacturer (industrialist) of the machine and production department utilize machines in their production line but both play a key role in predicting maintenance models. In this case, the manufacturer and the production department can keep the unique information about the same machine via unique instance space. Nevertheless, the industrialist provides wide information about technical description, design procedure, and experimental results. The production department maintains real-time data, evaluation metrics of the machine, and environmental monitoring of the machine (variant attribute space). The VFL offers different entities, and the accessible resources are considered different. The comparison of VFL and HFL is exhibited in Table 2.2.

TABLE 2.2
Comparison Parameters of VFL and HFL

Parameters	VFL	HFL
Attributes	Different	General
Instances	General	Different
Scalability	Lower	Higher
Training	Locally on consumer devices	Global updates on devices
Power utilization	Lesser than HFL	Consumes more
Privacy and security	Data sharing with global server	Data sharing with local server
Communication cost	Higher	Lower

2.4.1.3 Transfer Federated Learning (TFL)

In transfer federated learning (TFL), the nodes engaged do not distribute the attribute or instance space. This scenario evolved in different contexts. For instance, e-commerce and banking in various areas are involved in enhancing the credit score system of the bank. In this case, the bank keeps the data about preceding loan details, credit card payments, and balance in the customer account. In the same way, the e-commerce holds the details regarding shopping behavior of the customers (variant attribute space). Although the e-commerce customers are not the same as the bank customers, the bank can get viability from the shopping nature of e-commerce customers. Generally, TFL involves transfer of knowledge among two entities, and the available possessions are estimated differently.

Another scenario depicts TFL was deployed among two different datasets but varies with instance ID and attribute space. Assume two industries composing ML models to determine industry objects that rely on different cameras which yield images of different dimensions and color changes. It can be implemented to address the complete sample and attribute space. Through adopting a set of congruous instances, the general exemplification shows two attributes are obtained and used to yield detection according to the information from a single source [16].

2.5 META-PARAMETER OPTIMIZATION (MPO)

The optimization can be done through experimental results during FL stages. The experiments are benefitted through accelerators to analyze many configurations to obtain hasten results. The allocated MPO assures feasible examination of various configurations through governing resources and communication. The motive of MPO is to provide higher accuracy through tuning optimal parameters, number of epochs, and appropriate selection of optimization algorithms. The generally utilized parameter settings for the centralized system may not be the correct choice for non-independent identical data in FL. The client executes the searching process to implement neural architecture in FL parameter setting [17].

In Reference [18], a technique of single-shot FL MPO is developed. Every client performs local search through MPO by adapting Bayesian optimization to compete analogously in meta-parameter space with lower communication overhead. Further, the server consolidates the local surface and rebuilds the system to detect best parameters to the adapted populations. Thus, FL is initiated to train aggregate models by utilizing the best meta-parameter candidate solution. In References [19] and [20], a FL model to describe simulation settings to employ numerous FL configurations analogous over the candidate population of diverse devices is implemented which outperforms distributed MPO through exploration of searching models such as random search and grid search optimization.

Algorithm 1 – Meta-parameter Optimization Algorithm

Input: Patient's dataset $(d_1, d_2, \ldots, d_n)$, Number of epochs (M), Set of meta-parameter (Q)

Output: Optimal meta-parameter h_{optimal} for clients

Begin

Initialize the optimal meta-parameter h_{optimal};

Initiate optimal performance $P_{\text{optimal}} = 0$

  for $i = 1$ *to* M **do**

    Instance meta-parameter q from Q

    Broadcast q to all patients

    for every patient k **do**

      Train local model x_m using d_k and q

      Examine x_m on local set to yield optimal performance P_k

      Transfer P_k to the server

    end for

    Accumulate patient's performance $P = \{p_1, p_2, \ldots, p_n\}$

    Estimate average performance $\tilde{P} = \frac{1}{N} \sum_{k=1}^{N} P_k$

    If $\tilde{P} > \mathrm{P}_{\text{optimal}}$ **then**

      Update optimal meta-parameter $h_{\text{optimal}} = h$

      Update optimal performance $P_{\text{optimal}} = \tilde{P}$

    end if

  end for

Return h_{optimal};

TABLE 2.3
Comparative Analysis of Different PFL Optimization Models

Reference	Methodology	Contribution	Benefits	Datasets	Accuracy
[21]	HyperFed	Global sharing procedure is followed. It learns general attributes from various organization, and enhances performance	Outperforms CT reconstruction than other models	NIH-AAPM-Mayo	95.2%
[22]	HPN	Provides privacy protection for distinct clients	Articulates heterogeniety among clients	CIFAR-10	74%
[23]	pFedHN	Represents the constraints of parameter setting and personalization	Potent customization of models	CIFAR10	68.15%
[23]	pFedHN-PC	The centralized hypernetwork is trained to formulate set of models where each client has distinct model	Hypernetwork enables efficient amalgamation to models	CIFAR100 Omniglot	92.47%
[24]	HNTROJ	The backdoor-resistant algorithm is proposed to train the model.	Mitigates data poisoning and system replacement attacks	CIFAR10 FMNIST 100	85.4%

The comparative analysis of different personalized FL optimization models is exhibited in Table 2.3. The authors developed a model to address non-distributed data and client heterogeneity for parameter setting of FL. It also examines the efficacy of two FL algorithms such as FedAvg and FedProx. It performs splitting which consists of various forms of cancer namely lung cancer, colon cancer, and cervical cancer. It investigates the effect of meta-parameters in FL which results in specific challenges due to the disseminated behavior of the learning process. To obtain an optimal meta-parameter, Bayesian optimization method is utilized to tune the meta-parameters which improves the performance of the model.

A novel technique for individualized CT imaging is developed in FL, which is named as HyperFed. The HyperFed implements the idea of degrading the optimization problem within clinical organization into two parts: the global CT imaging and the local data adaption constraints. This ability of HyperFed handles these aspects by combining a global sharing network with an organization-specific hypernetwork. The global network provides stability and general features from distinct organizations. The organization-oriented hypernetwork yields meta-parameters to build the global network for specialized local networks [21].

To mitigate the constraints of personalized FL, the hyperparameter network (HPN) is introduced in the system. This model has the ability of computing hyperparameters through client encodings. It prompts training ML models for different clients through individual data distributions. It represents the constraints behind managing data distribution and reducing the communication costs across aggregate model training [22].

In Reference [23], personalized federated hypernetworks (pFedHN) have been presented to leverage hypernetworks. It is a centralized hypernetwork method which is used to train and yield user-specific approaches and also training every approach to desired features of individual users. The MPO searching and their optimization is delineated further. In MPO searching, the term (Q) illustrates the meta-parameter set which is modified and optimized amidst the tuning stages. From the approach, the optimal meta-parameters h_{optimal} are sampled from the searching space (Q) at every epoch. It intends to determine optimal parameter h_{optimal} which results in the finest performance of all contributing clients.

2.5.1 Optimization Interconnection

The interconnection of optimization that thrives through the number of epochs is specified as (M). It governs the number of epochs of meta-parameter optimization that will be executed. The following procedures are implemented: the meta-parameter (Q) is erratically sampled from searching space where the meta-parameters are disseminated to all clients. Every client trained its model by utilizing its associated dataset $\left(d_n\right)$ and sampled meta-parameter (Q). The performance of every model $\left(p_n\right)$ is calculated through a validation set. The performance of each client is transferred to the server, whereas the server accumulates the client's performance to compute average performance of the model $(\widetilde{P})$. When the average performance $(\widetilde{P})$ outreaches the optimal performance $\left(P_{\text{optimal}}\right)$, the finest meta-parameter h_{optimal} and finest performance $\left(P_{\text{optimal}}\right)$ are updated. The algorithm reiterates this procedure for desired number of epochs (M) to search and update the optimal meta-parameters. The meta-parameter optimization algorithm yields the finest optimal value for all participating patients.

In Reference [24], the objective was to investigate poisoning attacks in order to find and research previously hidden backdoors related to HyperNet-based PFL (HyperNetFL). Expanding on these discoveries, they unveiled HNTROJ, a model transferring attack that was the first of its type. Throughout the training phase, persistent malicious local gradients calculated across captured clients are effectively leveraged. The HNTROJ aims to reduce the number of clients needed for an efficient attack while sustaining anonymity to prevent discernible deterioration in the model's effectiveness on data samples.

2.6 ADVANCED MODEL OPTIMIZATION ALGORITHMS

Developing sophisticated model optimization strategies is crucial to improving ML models' performance. Here are a few significant methods and algorithms:

- Meta-learning in a federated environment
- Multi-task federated learning
- Reinforcement learning-based optimization

2.6.1 Meta-learning in Federated Environment (MLFE)

The objective of exploring the area in meta-learning is to enhance learning methods through exposition to a variety of tasks and datasets. The goal is to make it feasible for the method to find out functions quickly and accurately. In this area, the optimizer-based meta-learning using procedures such as Reptile and model-agnostic meta-learning (MAML) are followed. MAML is renowned due to its ability to quickly adjust to new and varied activities and generalize effectively. Most of the studies have demonstrated the efficacy of optimizer-based procedures such as MAML and Reptile in enhancing the ability to learn models [25]. These algorithms enable prompt compliance to novel functions and quickly generalize among a variety of datasets. The summary of contributions to meta-learning-based PFL model optimization strategies is shown in Table 2.4.

In Reference [26], the similarities between their ideas in order to demonstrate a connection between FL and meta-learning are highlighted. Specifically, they provide a comparison between several FL processes and the stages of meta-learning, especially meta-training and meta-testing. In order to improve a model's learning skills, it is trained on a variety of tasks during the meta-training stage of meta-learning. The researchers compare the previously mentioned stage to the FL universal training procedure, which uses data from several clients to build a global model.

TABLE 2.4
Contributions to Meta-learning-based PFL Model Optimization Strategies

Reference	Methodology	Highlights	Merits	Datasets	Accuracy
[25]	MAML	Measuring as a meta-learning technique is used to investigate the real-world uses of MAML in FL	Improved personalized accuracy indicates the necessity for a novel approach for predicting the generalization of personalized models	EMINIST	90.28%
[26]	FedHAM	PFL uses an hierarchical attention-enhanced meta-learning network. This network evaluates client similarities using a meta-learning approach	Successfully provides stability among the distinct characteristics of each client and their commonalities	V-MNIST	85.58%
[27]	G-FML	A system for cluster-based federated meta-learning that dynamically arranges clients into groups according to their data distributions	This approach enables tailored learning inside every cluster, where customized prototypes are developed through a learning strategy, allowing for more effective adaptation	FEMNIST	90.12%
[28]	Per-FedAvg	A customized kind of federated learning intended to pinpoint an initially shared model that customers may easily modify to fit their own datasets	This approach harnesses the advantages of federated learning while facilitating the development of personalized models	MINIST	79.85%
[29]	$P_{Fed}ML$	It decouples the optimizer of PFL models from the universal network, resulting in enhanced convergence rates. The procedure demonstrates a quadratic quicken for a heavily convex policy	Achieves higher rate of convergence	MINIST	95.62%
[30]	ARUBA	Incorporating function similarity with online convex optimizer and sequence techniques to design and analyze different methods	Enhances meta-test while conducting on basic standard issues	Omniglot	99.29%

Algorithm 2 – Meta-learning for PFL

Input: Set of clients K, Training data T_{meta}

Output: Trained model parameters φ

Algorithm:

Initiate meta model variables φ
for meta-iteration = 1 to $M_{meta\text{-}iterations}$ **do**
  for every client $k \in K$ do
    Sample task-specific data T_k from client k
    Initialize client-specific model variables $Ø_k \leftarrow φ$
    for local iteration = 1 to $M_{local\text{-}iterations}$ **do**
      Estimate loss L_k ($Ø_k$,T_c)
      Update client-specific model variables
  end for
  Updating meta-model variables
 end for
end for

The main goal is to enhance the universal prototype analysis for a variety of clients and tasks. Similar to this, the meta-testing stage of meta-learning comprises a less gradient descent procedure through swiftly adapting the prototype to new functions. It relates this stage to the FL personalization procedure, in which each client's local adaptation phase involves additional fine-tuning about the universal prototype based on local data [27]. The objective is to customize the universal prototype to every user's unique requirements and data distribution. The possibility for incorporating meta-learning techniques by enhancing the FL structure was highlighted. The procedures input subsist about the meta-trained data (T_{meta}) then the set of clients (k). The algorithm's output φ is the meta-trained prototype variables. Outcome of the method is presented by the final output. The meta-learning for PFL is demonstrated in Algorithm 2.

The Per-FedAvg approach is expanded upon by a personalized federated meta-learning (P_{Fed}ML). The main goal of P_{Fed}ML is to maintain a balance among FL customized and generalized performance. It is involved by adding a 12-norm legalization failure term to the optimized aim. Controlling the trade-off between generalization and customization is made possible by this normalization term. By modifying the strength in the normalization, the model's ability to be personalized for particular consumers while preserving generalization between clients can be managed [28].

The P_{Fed}ML algorithm outperforms the federal average and Per-FedAvg approaches in terms of accuracy and convergence. It makes use of the idea of Moreau envelopes which offer a continuous similarity of the regularizing. In federated meta-learning contexts, this improves performance and makes optimization more efficient. The P_{Fed}ML improves upon the Per-FedAvg method by utilizing Moreau envelopes and adding a l2-norm regularization loss. Better control over the ratio of generalization to personalization is made possible by this formulation, which enhances accuracy and convergence [29].

The ARUBA framework was presented and focuses on using online learning strategies to achieve flexible meta-learning in the context of FL. Introducing ARUBA to the federal average algorithm improves the model's performance in generalization and avoids the need for hyperparameter altering during the personalizing phase. A technique that shows promise for enhancing meta-learning's performance and adaptability in FL settings is the ARUBA framework [30].

2.6.2 Multi-task Federated Learning (MTFL)

The goal of MTFL is to complete more tasks with a single model rather than requiring a different model for every activity. Three types of MTFL can be distinguished in the context of object detection. Under the initial category, the total number of tasks required is represented by the number of head models. Hard parameter sharing occurs when the head models share a backbone. Every task has its own backbone when it comes to soft parameter exchange. Hard parameter sharing examples are to be located. In UAV applications, the joint segmentation and detection model includes a head model for lane line detection in the context of self-driving automobile applications [31].

The original detection model requires just a few modifications to enable multitasking. The goal is to turn the distance-prediction work into a classification task and then combine it with the vehicle classification task to create a single task. Next, some models use a number of auxiliary activities to enhance their primary tasks. To improve the object-detection model's learning process, for instance, three auxiliary tasks have been performed: (i) closeness labels, (ii) multi-object labels, and (iii) foreground labels.

The data availability demonstrates the efficiency of the MTFL solutions. In a scenario including extensive disaster monitoring such an assumption is not valid. This is still more to learn about how well MTFL can be trained on local devices using distributed datasets [32]. Additionally, the majority of these investigations use an empirical method to identify the ideal branching settings by conducting transfer learning on many combinations and then selecting the best one. Because of the additional instruction on every combination to assess the learning transfer performance, this method necessitates a lot of computing.

2.6.3 Reinforcement Learning (RL)

In the reinforcement learning (RL) model, a self-learning agent's aim is to maximize the value of a reward. RL models perform better than a lot of deep learning methods. RL deals with algorithms that are primarily used in games such as AlphaGo. The basic principle of RL is that an agent resides in a stochastic unknown environment. The agent functions in the sequential environment of the Markov decision process are a systematically regulated environment.

In a RL model, an autonomous agent's objective is to maximize a monetary reward. RL models perform better than a lot of deep learning methods in numerous applications [33–35]. The participants in RL understand the condition of the environment and as an outcome it takes an action that alters the situation in order to optimize a specific advantage. The agent must learn the relationship between the action and the subsequent change in reward, since the environment is unknown. In basic terms, this can be accomplished by understanding the implications of every action from trial and error.

The agent finally learns through many iterations of what to do in a given state for the greatest benefit. The policy and the value function are two of the sub-elements of a RL scheme. The policy defines the agent's actions in a given situation. Large and dynamic data issues can be solved by the RL model. In addition, the agent must act in every iteration where the RL can use prediction problems to solve NP-hard problems.

Policy gradients are one of the most convenient methods for training an RL policy network. Here, a random network is used as the starting point and an agent frame that generates a random output activity is then fed into the network. After receiving the output, the agent generates the subsequent frame. The procedure keeps going until a passably sound solution is found. In this case, the network may be a fully connected network. The agent randomly explores the environment and finds a better solution where the network's output must be sampled during training to prevent the action from being repeated indefinitely. When there is a clear reward signal for behaviors that can be performed in the environment where the agent is trained, the RL is more effective in real-world applications.

2.7 CONCLUSION

This chapter augmented by driving the taxonomy for centralized and distributed settings. It enhanced the distinction among centralized and distributed learning. The chapter explored different attributes of FL systems and challenges of the taxonomy. Additionally, PFL discussed the related technologies and model optimization techniques. It delineated different aspects of advanced optimization techniques and indicated the conflicts and opportunities in the domain of PFL. To provide secured services among various applications and instigate a stable environment, training of artificial intelligence models was presented. Through confronting these perspectives, the deployment strategies, taxonomy, and various scenarios of PFL methodology were realized and examined. It emphasized the current research in presenting the challenges and requires further improvements in different perspectives of PFL to assist future research in these areas.

REFERENCES

1. Mahato, G. K., Banerjee, A., Chakraborty, S. K., Xiao-Zhi, G., Privacy preserving verifiable federated learning scheme using blockchain and homomorphic encryption, *Applied Soft Computing*, 167, 2024. https://doi.org/10.1016/j.asoc.2024.112405.
2. Tanzib, H., Mushfiqur, M., Yeasin, R., Privacy preserving machine learning with federated personalized learning in artificially generated environment, *IEEE Open Journal of the Computer Society*, 5, 694–704, 2024.
3. Sekar, J., Aruchamy, P., A novel approach for heart disease prediction using hybridized AITH^{2}O algorithm and SANFIS classifier, *Network: Computation in Neural Systems, 36*, 1–39, 2024.
4. Yange, C., Lei, L., Yuan, P., A privacy-preserving federated learning framework with lightweight and fair in IoT, *IEEE Transactions on Network and Service Management*, 21, 5843–5858, 2024.
5. Sudha, M., Mahesh, R. V., Optimizing intrusion detection systems using parallel metric learning, *Computers and Electrical Engineering*, 110, 1–15, 2023.
6. Zaobo, H., Lintao, W., Zhipeng, C., Clustered federated learning with adaptive local differential privacy on heterogeneous IoT data, *IEEE Internet of Things Journal*, 11, 137–146, 2024.
7. Lathika, V., Jayachitra, S., Arunkumar, S., Balasubramani, M., Kalaichelvi, V., Secure multi-party computation for collaborative data analysis in network security, *International Conference on Intelligent and Innovative Technologies in Computing, Electrical and Electronics (IITCEE)*, 1–5, 2024. doi 10.1109/IITCEE59897.2024.10467913.
8. Schoinas, I., Triantafyllou, A., Ioannidis, D., Federated learning: Challenges, SoTA, performance improvements and application domains, *IEEE Open Journal of the Communications Society*, 5, 5933–6017, 2024.
9. Li, Q., Wen, Z., Wu, Z., Hu, S., A survey on federated learning systems: Vision, hype and reality for data privacy and protection, *IEEE Transactions on Knowledge and Data Engineering*, 35, 3347–3366, 2023.
10. Jayachitra, S., Devi, S. N., Hariprasath, S., Khan, J. A., Feature extraction using Levy distribution-based salp swarm algorithm, *Lecture Notes in Electrical Engineering*, 1146, 165–175, 2024.
11. Liu, Y., Liu, Y., Huang, A., Luo, Y., Fed vision: An online visual object detection platform powered by federated learning, *Proceedings of the AAAI Conference on Artificial Intelligence*, 34, 13172–13179, 2020.
12. Zhou, X., Ke, Y., Cui, Q., Stfl: spatio-temporal federated learning for vehicle trajectory prediction, *IEEE International Conference on Digital Twins*, 1–6, 2022. doi. 10.1109/DTPI54763.2022.00010.
13. Zeng, T., Semiari, O., Mozaffari, M., Chen, M., Saad, W., Bennis, M., Federated learning in the sky: Joint power allocation and scheduling with UAV swarms, *Proceedings of IEEE International Conference On Communication (ICC)*, 1–6, 2020. doi. 10.1109/ICC40277.2020.9148776.
14. Wan, S., Lu, J., Fan, P., Shao, Y. How global observation works in federated learning. Integrating vertical training into horizontal federated learning, *IEEE Internet of Things Journal*, 10, 9482–9497, 2023.

15. Zhang, L., Li, A., Peng, H., Han, F., Huang, F., Li, X. -Y., Privacy-preserving data selection for horizontal and vertical federated learning, *IEEE Transactions on Parallel and Distributed Systems*, 35, 2054–2068, 2024.
16. Chen, J., Li, J., Huang, R., Yue, K., Chen, Z., Li, W., Federated transfer learning for bearing fault diagnosis with discrepancy-based weighted federated averaging, *IEEE Transactions on Instrumentation and Measurement*, 71, 1–11, 2022.
17. Subramanian, M., Rajasekar, V. R., Effectiveness of decentralized federated learning algorithms in healthcare: A case study on cancer classification. *Electronics*, 11, 1–14, 2022.
18. Xu, C., Qu, Y., Eklund, P.W., Xiang, Y., Gao, L., Bafl: An efficient blockchain based asynchronous federated learning framework, *IEEE Symposium on Computers and Communications (ISCC), IEEE*, 1–6, 2021. https://doi.org/10.1109/ISCC53001.2021.9631405.
19. Chen, Y., Wang, L., MetaFed: Federated learning among federations with cyclic knowledge distillation for personalized healthcare, *IEEE Transactions on Neural Networks and Learning Systems*, 35 (11), 16671–16682, 2024.
20. Fahad, S., Yuwen, C., Model optimization techniques in personalized federated learning: A survey, *Expert Systems with Applications*, 243, 122874: 1–19, 2024.
21. Yang, Z., Wenjun, X., Hypernetwork-based personalized federated learning for multi-institutional CT imaging, *IEEE Transactions on Neural Networks and Learning Systems*, 36 (2), 3136–3150, 2025.
22. Cheng, A., Wang, Z., Yaliang, L., Cheng, J., HPN: Personalized federated hyperparameter optimization, *Arxiv*, 1–10. doi https://doi.org/10.48550/arXiv.2304.05195.
23. Shamsian, A., Navon, A., Fetaya, E., Chechik, G., Personalized federated learning using hypernetworks, *International Conference on Machine Learning*, 139, 9489–9502, 2021.
24. Phung, L., Nhathai, P., Model transferring attacks to backdoor hyper network in personalized federated learning, *arxiv*, 2022.
25. Jiang, Y., Konecný, J., Rush, K., Improving federated learning personalization via model agnostic meta learning, *arxiv*, 2019.
26. Gao, Y., Wang, P., Liu, L., Configure your federation: Hierarchical attention-enhanced meta-learning network for personalized federated learning, *ACM Transactions on Intelligent Systems and Technology*, 14, 1–24, 2023.
27. Yang, J., Shi, R., MedMNIST v2-A large-scale lightweight benchmark for 2D and 3Dbiomedical image classification, *Scientific Data*, 10, 41, 2023.
28. Fallah, A., Mokhtari, A., Personalized federated learning with theoretical guarantees: A model-agnostic meta-learning approach, *Neural Information Processing Systems*, 33, 3557–3568, 2020.
29. Dinh, T., Thanh, T., Tran, N., A new look and convergence rate of federated multitask learning with Laplacian regularization, *IEEE Transactions on Neural Networks and Learning Systems*, 35, 8075–8085, 2021.
30. Khodak, M., Balcan, M., Talwalkar, A., Adaptive gradient-based meta-learning methods, *Neural Information Processing Systems*, 32, 5915–5926, 2021.
31. Qian, Y., Dolan, J. M., Yang, M., DLT-Net: Joint detection of drivable areas, lane lines, and traffic objects, *IEEE Transactions on Intelligent Transportation Systems*, 21, 4670–4679, 2020.
32. Chen, Y., Zhao, D., Lv, L., Zhang, Q., Multi-task learning for dangerous object detection in autonomous driving, *Information Science*, 432, 559–571, 2018.

33. Forster, A., Machine learning techniques applied to wireless ad-hoc net- works: Guide and survey, *3rd International Conference on Intelligent Sensors, Sensor Networks and Information*, 365–370, 2007.
34. Mismar, F. B., Choi, J., Evans, B. L., A framework for automated cellular network tuning with reinforcement learning, *IEEE Transactions on Communication*, 67, 7152–7167, 2019.
35. Sun, Y., Peng, M., Mao, S., Deep reinforcement learning-based model selection and resource management for green fog radio access networks, *IEEE Internet Things*, 6, 1960–1971, 2019.

3 Automation of AI and IoT-Based Data-Driven Decision-Making Approaches Using Federated Learning Systems

M. Sundarrajan, Mani Deepak Choudhry, Rajesh Kumar Dhanaraj, and Mariya Ouaissa

3.1 INTRODUCTION

The advancement and widespread adoption of billions of mobile and internet of things (IoT) devices have become evident in recent years. This rapid growth of IoT sensors, devices, objects, and resources has placed significant pressure on the current IoT and distributed sensor network infrastructures. Emerging technologies and intelligent approaches have a vital role to play in addressing these increasing demands. This has led to the creation of a new interdisciplinary domain and paradigm, the artificial intelligence internet of things (AIoT) [1]. AIoT has started to attract substantial attention from both research communities and industries alike.

The broad acceptance and integration of artificial intelligence (AI) technologies have led to the rise of more AIoT applications. These applications typically have high computational demands as they must run machine learning (ML) or deep learning (DL) algorithms while meeting real-time processing requirements. However, IoT devices are limited in terms of resources, such as computational power, storage capacity, and communication abilities, which heightens the challenges associated with designing and deploying solutions that satisfy the quality-of-service needs of AIoT and its related applications.

During the last few years, IoT has been advancing rapidly and promises ubiquitous sensing and computing in connecting heterogeneous collections of devices to the internet [2]. These ubiquitous IoT devices create enormous amounts of data, and with the use of AI techniques, particularly DL, data models have been developed that

DOI: 10.1201/9781003532323-3

facilitate smart IoT applications, including intelligent healthcare, intelligent transportation, and intelligent cities [3]. Traditionally, AI functions were implemented on cloud servers or data centers for training and modeling [4]. However, this approach poses considerable challenges with the data explosion from IoT.

Exponential increases in IoT devices as well as new applications generate tremendous data traffic daily. This creates huge demands on the internet infrastructure. Substantial investments in internet infrastructure are required to raise its capacity. Big data analytics and AI techniques, including DL and ML, are well advanced, making these collected data useful for many applications. AI has been recently widely applied across all communication disciplines for the past few years [5]. As an example, ML algorithms are used in the field of antenna selection in multiple-antenna wireless systems. DL methodologies have also been recently used to mitigate computation offloading challenges in IoT settings by the use of edge computing [6]. Other examples of deep reinforcement learning (DRL) enhancement of resource allocation at network edges include solutions to traffic classification and edge caching, network security, and offloading data.

However, all traditional AI models require processing of data in a central location, making users download their local data to upload it to the central location for model training. As a result, this strategy raises many serious privacy problems because most individuals want to be in complete control of their local data and do not want the central entity to access these data. Also, demanding a central cloud infrastructure with high computing capabilities and significant storage for model training is crucial [7].

In sharp contrast, current advancements in computing technology and the high integration of smart devices in daily life reveal that many IoT devices today are installed with computational as well as storage capacities similar to those of a desktop computer ten years ago [8]. Consequently, it is quite inappropriate to use classical ML models in large-scale IoT networks and does not make full use of computing resources. This calls for a new learning paradigm where the training data never needs to be centralized from individual IoT devices.

In this regard, the emergence of federated learning (FL) has been a promising step toward developing intelligent and privacy-conscious IoT systems. Within this context, FL is defined as a decentralized approach to collaborative AI in which it becomes possible to train models without the need for raw datasets to be redistributed by coordinating multiple devices with the aid of a central server [9].

In an intelligent IoT network, different types of IoT devices act as participants that collaborate with an aggregator such as a server for neural network training [10]. First, the aggregator initializes a global model with some learning parameters. The devices then download the current model from the aggregator, compute their local updates concerning their dataset using methods such as stochastic gradient descent, and send them back to the aggregator. The aggregator aggregates the local updates together into a new improved global model. This mechanism, apart from improving the quality of the training of the model, also reduces the exposure of risks associated with the private data of the users. The cycle continues as local devices download updated global models for them to compute their next local updates until the global process reaches completeness. FL offers various potential benefits such as:

- Enhancing data privacy
- Reduced latency communications
- Learning quality enhancement

Based on these unique advantages, FL has been proposed for an extensive set of IoT applications, such as smart healthcare, smart transportation, and unmanned aerial vehicles (UAVs). For instance, FL was capable of constructing smart health services in which it could train the ML models without sharing any type of patient data among different hospitals. Therefore, through health data owners such as hospitals, the above result will enable these parties not to share their healthcare records directly but instead to train AI models locally and upload only trained parameters to the aggregator for computing the global model. Therefore, this shall promote collaborative healthcare environments across various hospitals to optimize the speed of patients' diagnosis and treatment without trading off privacy.

FL has been experimented to provide smart vehicular services in transportation systems, including the monitoring facility of smart vehicles, autonomous driving, and forecasting road safety with vehicle detection. These are achieved by coordinating the vehicles with roadside units for collaborative learning of data. The recent success of FL-IoT applications shows the importance currently given to this newer research area and thus it becomes a proper subject to further explore and develop.

3.2 RELATED WORKS

This section presents a comprehensive analysis of applications of FL across a broad spectrum of IoT use cases: smart healthcare, smart transportation, UAVs, smart cities, and smart industries. The section also comes with some indispensable critical insights into the survey of these applications. Further, the chapter discusses several significant research challenges and sets forth interesting future directions in FL for IoT settings. In summary, the contributions of this chapter are as follows:

- This chapter is an effort to critically review FL in IoT networks, initiating from the discussion of the recent achievements in FL and IoT-related technologies and the motivation behind their integration. In such core IoT services as data sharing in IoT, offloading and caching, detection and localization of attacks, mobile crowdsensing, privacy, and security in IoT, the opportunities that FL brings to such areas are discussed.
- A comprehensive analysis of FL applications in various IoT domains will be conducted, including sectors such as smart health, smart transportation, UAVs, smart cities, and smart industries. The analysis also involves communication and networking implications with each of these applications, with taxonomy tables categorizing the primary technical characteristics, contributions, and limitations of every FL approach applied in an IoT context.
- The following steps summarize the drawing of key insights from this analysis of FL in IoT services and applications. In its conclusion, the chapter identifies

critical challenges to further research directions in unlocking the real potential of FL within IoT networks.

A large number of recent studies have been seen focusing on the integration of FL and the IoT. FL's main philosophies were analyzed in depth, including an investigation into the critical elements of the aforementioned technology, such as data distribution, models of ML algorithms, privacy-preserving schemes, and frameworks designed to communicate between different nodes [11]. Similarly, FL has been approached from an architectural standpoint, and then from that perspective, indicated a few of the most basic applications of FL within different organizational structures [12].

The authors in References [13–15] discussed several architectures, software frameworks, platforms, and protocols of FL, and issues experienced during the implementation process. FL's data privacy and security vulnerabilities were in-depth analyzed in Reference [16]. Further, the recent applications of FL in mobile edge networks have been discussed by authors of Reference [17], where FL was revealed to help improve the efficiency of the edge network. FL has also been acknowledged in the field of wireless networks.

An overview of FL in fog radio access networks was presented in Reference [18] with model accuracy correction and model compression as two important concepts. Another comprehensive analysis presented the various applications of 5G in FL, which include edge computing, spectrum sharing, and core network optimization [19]. Further, the integration of FL into future 6G wireless systems has been studied by considering the difficulties raised by such a transition [20]. The methods for FL training that preserve privacy were discussed by authors of Reference [21], and an overview of leading FL model architectures and their use cases in health informatics was provided in Reference [22]. On the other hand, Reference [23] discussed the potential of FL in vehicular networks, and FL–UAV models with applications and their representative models in the real world were introduced and discussed in Reference [24]. Such diversification of work depicts the broad use cases and ongoing developments in FL across the horizon of IoT and wireless communications. Table 3.1 gives an overview of these analyses.

3.3 FEDERATED LEARNING SYSTEMS AND DECISION-MAKING APPROACHES

FL, since its early appearance in 2016 [25], has promised great potential for shifting intelligent IoT applications with a new kind of innovative AI approach, due to their decentralized privacy-focused approach. Today, this novel form of distributed AI is expected to lead to a fresh wave of intelligent IoT systems, especially those built upon more advanced architectures of FL. With recent improvements in hardware for mobile devices and heightened concerns over privacy, FL is an ideal solution for IoT systems distributed in nature with the ability to perform data training as part of the AI processing directly at the network edge on IoT devices. This decentralized approach ensures that the user's data remains local, allowing for collaborative training of a

TABLE 3.1
Overview of Existing Analysis on FL-Related Topics

Reference	Concept Involved	Performance Metrics Used	Limitations
[11]	Overview of FL and key system components	Model accuracy and privacy levels	Lacks detailed discussion on specific applications
[12]	Architectural perspective of FL	Accuracy and training speed	Limited focus on practical implementations in business
[13–15]	FL architectures, software, platforms, and protocols	Computational cost and model efficiency	May not cover all emerging research challenges
[16]	Technical challenges in FL systems	Data privacy levels, attack resistance	Primarily focuses on bottlenecks, less on solutions
[17]	FL in mobile edge networks	Latency, resource utilization	Challenges of implementation are highlighted but not fully explored
[18]	FL in fog radio access networks	Model accuracy, compression rate	Discussion on accuracy loss correction may be too theoretical
[19]	FL for wireless communications	Network throughput, latency, accuracy	Roles in 5G applications were discussed, but limited in depth
[20]	FL in 6G wireless applications	Model accuracy, network efficiency	Highlights challenges but lacks detailed mitigation strategies
[21]	Data preservation methods in FL	Privacy preservation level, model accuracy	May not address all aspects of user protection
[22]	Key architectures of FL models	Model accuracy, data sensitivity	A brief introduction to health informatics lacks depth
[23]	FL in vehicular networks	Detection accuracy, latency, privacy	Limited exploration of practical applications
[24]	Integrated FL–UAV models	Task completion rate, energy consumption	Representative use cases may not cover all scenarios

global model without sharing such data with third parties. This method improves privacy for IoT users while maximizing network resources for service providers, making FL an attractive alternative to the traditional centralized AI models. It allows the wide-scale deployment of diverse IoT services and applications [11].

Figure 3.1 demonstrates an overview of FL in IoT systems. Usually, each IoT device functions on its local dataset, and it might have an aggregation server located

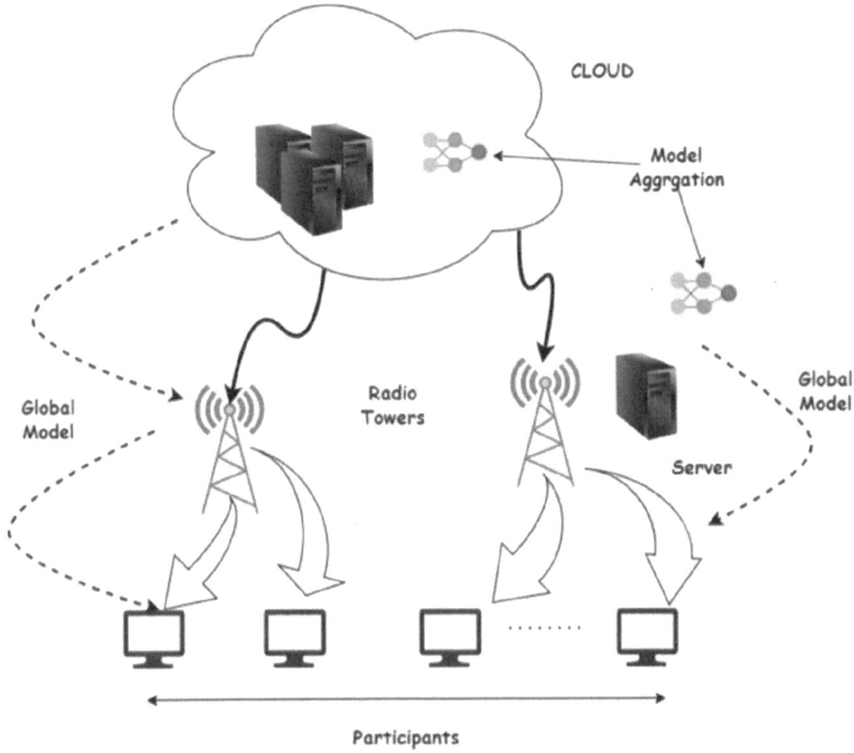

FIGURE 3.1 General FL–IoT framework.

at the network edge or within a remote cloud computing environment. Depending on specific parameters, each FL model offers distinct advantages and limitations. FL models with a server placed at the network edge are best suited to applications that require low latency, location awareness, or contextual insights from the network, thus fully exploiting the potential of FL for efficient, responsive, and context-driven IoT solutions. On the contrary, cloud-based FL should be taken up for applications in which large numbers of IoT devices across multiple regions have to be involved and require robust computing and storage capabilities.

3.3.1 Fundamentals of FL

Digital data has exploded to gigantic proportions in recent times, supported by various forms of information technology applications, including IoT, smart cities, and Industry 4.0, and rapid growth in digitization. Conventional ML approaches that rely on training algorithms on the data located either on a local hard disk of personal computers or at central locations could not serve well for real-time applications with this generation of big data. Scalability along with the time complexity is becoming a limiting factor in their efficiency. Large-sized training datasets make the ML algorithm insufficiently trained within the constraint of memory and computations. For this, data manipulation language (DML) comes into the scenario. In DML, algorithms of ML can be learned from the parts of distributed data across different nodes. While DML brings a number of benefits in numerous applications, it also brings some limitations. Major issues where DML lags include privacy and security concerns, reliance on platforms, absence of standardized evaluation measures for multi-domain adaptation (MDA) algorithms, and more.

The growth of smart devices that are enabled with measuring sensors has been rising pretty rapidly. These devices generate amounts of datasets which create significant opportunities for learning from intelligent systems. As the volume of data increases, so does the possibility of a system improving its accuracy. AI can be defined as a computational approach that allows machines to perform tasks in a manner that appears intelligent to humans. Knowledge acquisition, prediction, problem-solving capabilities, and a learning and implementation ability that involves reasoning, manipulation, and movement together characterize AI. AI systems can detect faults and perform assessments without any direct help of experts. An AI system can solve complex problems in an auto-efficient manner. It is commonly applied for the optimization of traditional technologies by using data-driven strategies. AI includes several subfields, such as ML, neural networks, and DL. ML is regarded as a subfield of AI whereby machines are made capable of learning from experiences, thereby improving with the passage of time, without an explicit program.

Industrial IoT (IIoT) essentially is a network of people, sensors, machines, and computers that enables smart operations in industries with the help of ML and analytics. IIoT means smart devices and connected sensors can collect data from manufacturing plants based on AI output predictions of results and automatic processing of outcomes. This technology has been a rejuvenator in industries also, which were found to lack high-skilled personnel. One of the most interesting strengths of IIoT is that it has minimal or zero dependence on human intervention. It allows the system to make intelligent decisions autonomously by learning from the sensor data. It gives way to the development of smart machines, obtaining hitherto impossible levels of accuracy due to human interference. As a disruptive technology, IIoT is going to transform product manufacturing and delivery, change factories into smart environments, and improve the safety of frontline workers in adverse conditions with increased security. All this will minimize human errors and efficiency while bringing huge cost savings.

FL is one mode of DML. FL is a unique approach wherein training of the models of ML algorithms takes place on data available on decentralized devices, such as those found in mobile phones and other smart devices. There is no need for any device participating in this process to share its samples. One can thus be assured of privacy and security. The basic principle of FL is training ML models on locally available data with periodic sharing of model parameters between nodes, which in this context happen to be devices. This process can learn a global ML model without exchanging the training data, which remains local on the devices. The global model is running on a central server that averages together the individual models' parameters after all the devices submit their updates. This collaborative learning allows the individual devices to still benefit from a shared model. Therefore, FL offers a set of advantages over many other traditional machine models, such as improved data privacy, reduced latency, and power saving. Besides protecting sensitive user information, FL could allow personal machine models to users and, therefore, eventually enhance their experience in interactions [12].

The key differences between FL and DML are mainly with the focus. DML mainly concentrates on parallelizing the power of computing, whereas FL emphasizes the training of locally developed datasets on devices that might include smartphones. In DML, datasets are assumed to be uniformly distributed (in size) across various nodes in the architecture. FL, however, supports different sizes of datasets at distinct nodes. Furthermore, the nodes in DML are primarily data centers that use the best communication technologies and robust computational capacity. On the other hand, FL nodes consist of mostly battery-dependent devices that rely on communications like Wi-Fi and cellular networks for communication; therefore, they cannot be relied upon and are connected to the network frequently.

3.3.2 Need of FL

The continuously increasing number of IoT resources and associated applications requires a huge volume of data processing. Big data analytics and computational methods, such as ML and DL, made it possible for users to process data effectively. AI applications are successfully utilized in the management of optimized resources and efficient antenna selection in wireless systems and several other disciplines of communication networks. Traditionally, AI models have to insist that users have to share individual data with the central network for learning purposes. Major problems with this type of approach are issues of sensitive user information privacy. FL effectively addresses these issues by allowing decision-making processes to depend significantly on information distributed across multiple training nodes while simultaneously ensuring privacy and security.

ML models have significant importance for enterprises, but traditional centralized ML methods suffer from severe drawbacks such as no support for continuous learning on edge devices and accumulation of sensitive data on centralized servers. FL can address these drawbacks by providing a method of model training that is decentralized and privacy-aware directly on edge devices. In traditional ML settings, a single model is built by using all available training data in a central place. Such an approach naturally works where there exists a central server that can provide instant predictions. However, in the context of mobile computing, users expect low latency and a delay in communication between a user's device and a far-central server can degrade the experience of the user.

One potential remedy involves the direct deployment of the model onto the device; however, this strategy introduces additional challenges, as it constrains the model's capacity to perpetually acquire knowledge from new data in the absence of access to the entire dataset. The main challenge in centralized ML is that the aggregation and possible retention of user information in a single repository would contravene privacy regulations and make it more likely for data breaches to occur. FL mitigates these challenges by facilitating ongoing learning on end-user devices without the necessity of sending user data to a central server.

By employing FL, models are collectively trained across various devices, thereby safeguarding data privacy while enhancing model accuracy and adaptability in real time, thus positioning it as a robust alternative to conventional centralized methodologies. Such models developed based on data from diverse sources may provide much insight and good predictive outcomes. However, raw data is difficult to transfer directly to a central system because it occupies high bandwidth usage, introduces security-related threats, and often exceeds storage capacity.

The distributed approach of FL optimizes learning rather than transferring raw data centrally for learning by using the accumulated raw data locally. FL is particularly helpful in the implementation of smart city initiatives. FL allows policymakers to extract sensitive information from IoT devices distributed throughout cities without breaching the privacy of individuals. This decentralized learning framework provides users with the ability to analyze shared insights without any access to personal information belonging to other participants. Once the training is over, the updated global model at the server is distributed back to each of its clients. Each of the client devices can download this refined model and use cloud distribution for evaluating model performances or interference on the specific devices. In this way, FL saves bandwidth, enhances data security, and holds up smart city development through effective, privacy-preserving data use.

3.3.3 Features of FL

FL, introduced by Google, allows for on-device learning while preserving the privacy of sensitive user information, optimizing distributed data processing, and strengthening privacy. This groundbreaking methodology would allow training ML models without consolidation of all user data samples in a centralized repository, thus promoting a resilient framework for distributed analysis, especially in the IIoT space. A salient feature of FL is the ability of a central server to supervise the training process and assess contributions from each participating device, an aspect that is particularly important in the IIoT, where efficient and resilient models should be kept in balance with rigorous data security requirements. So, despite FL's potential, challenges remain in its implementation in industrial settings, and it can be complicated regarding privacy and sensitive information security [14].

A notable application of FL is in aggregating the electronic health records from different hospitals, effectively allowing knowledge sharing across institutions while simultaneously preserving personal information. There exist three main types of FL: horizontal FL, which performs on the same feature space across devices; vertical FL, which aggregates datasets with different feature domains to train a global model; and federated transfer learning, where a pre-trained model, developed using a related dataset, is adapted to address a distinct problem different from the one it was originally designed for. Overall, FL represents a powerful paradigm for leveraging distributed data in a secure and privacy-preserving manner, making it particularly relevant in sectors where data sensitivity is paramount.

FL denotes a revolutionary step forward in the distributed ML area and explicitly emphasizes data privacy preservation. Recent research has witnessed an upward trend in distributing frameworks based on privacy, thereby enabling distributed collaboration among multiple geographically distributed nodes using a communication network controlled by a central server where each node participates in contributing toward a different aspect of a single task. Some of the primary characteristics of FL are as follows:

- **Universal Cross-Organization Scenario:** Originally invented by Google, FL is a distributed ML technology wherein participants develop a global model while retaining control over their local datasets. The concept has since been defined to encompass all decentralized ML methodologies to preserve privacy.
- **Highly Non-Identically Distributed:** FL contains highly distributed data that is dispersed over a vast number of edge nodes or devices. Every node likely contains data that is less comprehensive in comparison with the collective dataset accessible across all nodes, revealing the heterogeneity of data distribution.
- **Decentralization:** FL neither fully embodies the connotation of decentralization in the technical sense nor does it bear one single definitive central point of emphasis. Yet, each client goes in some way to influence the central model without being stuck as the central one. The parameter servers are more like central hubs, providing all resources to be capable of smoothly cooperating between the nodes.
- **Equitable Status of Every Node:** FL ensures that equal status is acceded to every participant. Thus, all nodes remain the same in the aspect of capability and opportunities in a system. Thus, though nodes with large datasets maintain more influence, the structure sustains a balanced collaborative environment at all times when creating the model.

These characteristics make FL compellingly suitable for applications where ML efficiency is mandated along with strictures of data privacy.

3.3.4 Motivation for Integration of FL with AI and IoT

3.3.4.1 FL with IoT

There are various challenges, including security, privacy, and communication costs in the rapid growth of applications in the IIoT. Combining FL with IIoT presents promising solutions to most of these challenges. This section outlines the key reasons for integrating FL with IIoT applications:

- **Protections on Security and Data Privacy:** While developing useful knowledge and identifying patterns from the data produced by IoT devices in an IIoT setting, it employs the applications of ML and DL algorithms [16]. Such algorithms require the functionalities of never-ending training over gigantic datasets coming from diversified industries and geographies for complex pattern detection. However, aggregating all the data sets into a central location to train the algorithm raises the threat of sensitive information being compromised by potential cyber threats. Using FL eliminates the need for transferring data to a central site to train the algorithm thereby keeping sensitive information safe and enhancing privacy.

- **Reduced Communication Costs:** Sensors put in place in IIoT systems generate humongous data that is accumulated onto edge devices for analytical purposes before uploading it onto centralized environments like the cloud. The huge transfer of such large datasets indeed comes with a massive communication cost. FL bypasses this problem as only aggregated results produced after applying ML/DL models are transmitted over to the cloud, thus lowering the expenses on communication.
- **Improved Network Utilization:** Applying FL enables an ML/DL model to process data from IIoT applications directly at the edge, not transmitting central places a summarized result set. Such an approach significantly reduces network loads and thus achieves better network efficiency overall.
- **Scalability at Scale:** FL applied in the context of IIoT is considered to make DL algorithms scalable because of the no necessity of training at local levels. The central algorithm is primarily trained on aggregated information and results from individual edge devices; therefore, this contributes to a drastic increase in the scalability of central learning algorithms and helps in effective learning from data obtained from different edge or local sources.

In normal FL, the process works on a centralized server. All the updates from both the ML and DL models are collected and then aggregated and relayed to the global model on the server. Security issues tend to crop up in the transfer of the updates to the server or in the distribution of the global model to the clients. To mitigate these security concerns, blockchain technology can be utilized for storing model updates in an immutable manner. IIoT devices train models with local data and then transmit their updates to an edge node as soon as the training is done for each device as shown in Figure 3.2. The updates that have been collected by each edge node are then forwarded to a fog node. As such, a fog node collects updates from all of the edge nodes and further condenses the data. These updates are then formatted into blocks, which are verified by a smart contract. Once validated, the block is added to the blockchain, so the block becomes tamper-proof and secure. Finally, all local updates from each of the edge nodes are summed up and passed to the centralized server for training the global model. After training, the updated global model will be sent out to all edge nodes again so that they can get in sync with the end devices connected to them.

3.3.4.2 FL with AI

FL, a decentralized collaborative AI framework, is especially beneficial for intelligent healthcare because it facilitates cooperation among different clients, such as

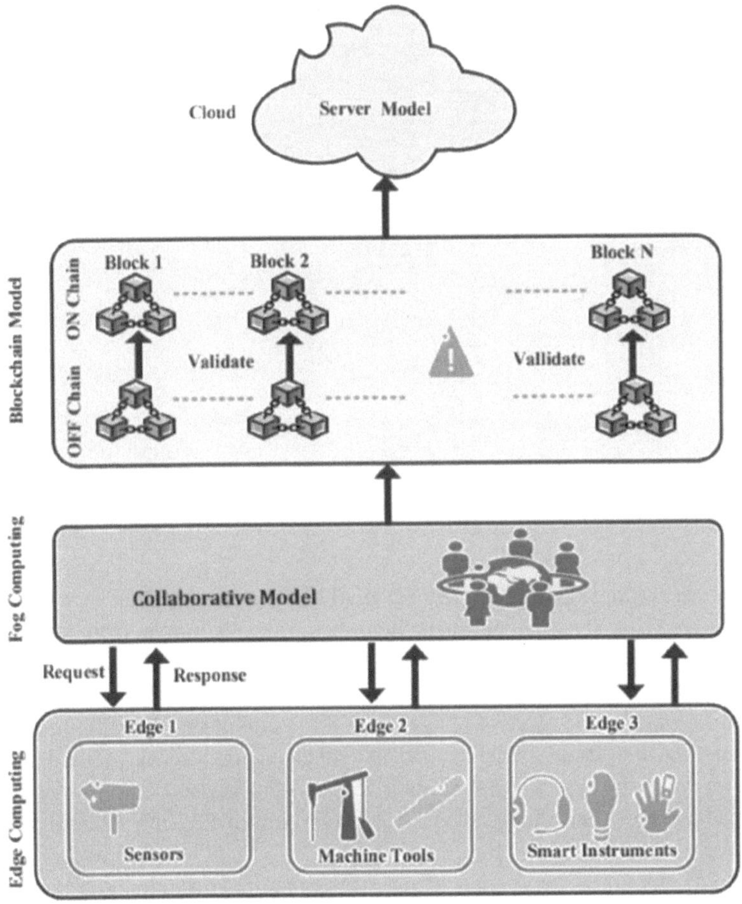

FIGURE 3.2 FL with IoT–blockchain.

hospitals, in AI training while eliminating the requirement to exchange local data. Figure 3.3 depicts the classification for FL alongside AI.

Real-time data gathering mainly due to the growth of technologies, such as ML and DL, has significantly driven improvements in autonomous control-based systems. IoT-based systems are more widely deployed; consequently, there is exponential data production by each sensor, and cloud storage solutions have made data management both highly accessible and efficient. However, a careful analysis of the entire process of data gathering and incorporation into advanced DL models is necessary for the development of AI-based frameworks. Should the actor manipulate data as it is transmitted from source to destination, then the reliability of predictions would most probably be severely compromised threatening information that is highly sensitive.

Under this paradigm, several techniques are implemented to enhance the security of data, out of which is FL. In FL, unlike transmitting the local data to the central

FIGURE 3.3 Taxonomy of FL with AI.

server for developing the model, instead, it allows the development of a model that is forwarded to each local client. Each client then trains the model from its local data and sends the updated parameters or weights back to the central server. Importantly, no actual data transfer takes place; only the results of local training are fed back into the central network. The server will aggregate these outcomes, and calibrate weights, before sending an updated model back to all connected clients. This method preserves locality in raw data, thus improving data security, but it also demonstrates the significant advantages that bring a successful integration of FL with AI across various applications.

FL represents a new generation of AI, which features decentralized training and learning processes. This architecture makes it possible to share the model across multiple devices without violating the privacy of individual users' data. This method efficiently ensures user privacy as it also enhances security for data by eliminating the use of raw data. Meanwhile, intensified concerns over user privacy also enhanced the role of FL, which operates in the intersection of AI, cloud technology, and user devices. It plays a vital role in creating more secure and robust infrastructures by integrating user equipment with the cloud in more intensive ways than the previous systems. Additionally, thanks to the properties of FL, it can deal with imbalanced and non-IID (independent and identically distributed) data without compromising the performance of the AI system involved. This helps larger devices in many contexts train with their local data and local models seamlessly.

3.4 DISCUSSION

3.4.1 Open Issues

The FL process is essentially an ongoing process wherein the model keeps getting updated over time. It brings along with it a host of challenges and issues that emerge as this concept is further developed. Researchers are challenged to solve these problems efficiently within a short span. Since the whole procedure operates over the internet, privacy and security are significant concerns, especially in advanced healthcare systems.

Comparatively with other data, health-related data is especially sensitive. Therefore, to ensure higher security, this type of data is being adapted into several technologies such as software-defined networking (SDN) and blockchain into healthcare systems [20]. However, FL is one of the advanced approaches of ML that needs massive datasets. Perhaps the biggest challenge related to the acquisition of appropriate datasets to be used in medical applications is the complexity of managing such datasets in the healthcare environment [21]. This is partly because such datasets often reside outside the local space in which the user operates, and distributed training is often conducted in this distributed space so that these users can modify a global model locally. Spatial separation may thus hinder efficient cooperation and the effective utilization of data. While health data may yield poor efficacy due to overdispersion in the available data across various domains and presence of outliers in instances, factors such as these lead to biased results that violate the reliability of the learnt models.

Additionally, processing health data and training ML models is both expensive and time-consuming, necessitating the development of more efficient methodologies. The healthcare industry is becoming progressively dependent on sophisticated ML methodologies; however, the logistical and financial challenges related to data management and model training may hamper further progress. Therefore, there is an emergent need for new approaches that support data governance, maintain data integrity and representativeness, and enhance the effectiveness of model training. This can include FL along with a myriad of decentralized methods to address privacy, access, and model robustness within data and may therefore improve ML efficiency in healthcare applications.

New methods and technologies have been explored by researchers to efficiently improve the performance of systems in the healthcare sector. Among those areas, security is still paramount as personal and sensitive patient data are required to be processed for the development of intelligent healthcare solutions that could be made accessible and effective anywhere with the help of the internet of medical things [23]. The adoption of SDN, blockchain, and network functions virtualization (NFV) can improve security. In the age of AI, data augmentation techniques can be used to synthesize the existing datasets to create more data. It is required since FL requires lots of data. Further analysis of data would be essential to ascertain optimal training times and maximize performance. The current models of ML are still largely behind generalization ability in medicine. The performance of convolutional neural networks (CNNs) can be significantly enhanced through restructuring and thorough testing. The models must also have cost analysis by use of statistical methods relative to their applications in the health industry.

Since there is sensitive and confidential information involved, data privacy can be considered one of the foremost crucial issues in IIoT that requires stricter security features than any other. Several data techniques have been used in detecting unauthorized users and stopping or curbing the attacks mainly regarding training of static as well as dynamic data streams of FL. There are two kinds of privacy, namely global and local, where global privacy is to access model updates mainly on the centralized server, while local privacy is to update to the server alone. The leakage of data becomes one of the major hurdles to effective data sharing, researchers have therefore focused more on preventing strategies that ensure safe data protection from any breach which would lead to financial loss and increased security costs. Thus, a distributed environment makes sure that secure data transmission is essential for the protection of sensitive data within IIoT systems.

To address these issues, blockchain technology is utilized in IIoT-based FL systems to ensure data integrity, facilitate adequate customer information, and enhance training capabilities. However, blockchain-based FL systems encounter challenges during the design phase, particularly regarding data heterogeneity in IIoT when it comes to early fault detection. Despite this, existing security mechanisms often fail to prioritize the development of effective frameworks and do not adequately address the reliability issues associated with blockchain technology.

3.5 RESEARCH DIRECTIONS

With the wide adoption of IoT-based solutions throughout industries, the generation of data is becoming at a rapid pace. This also forms a great challenge to data scientists as one of the major concerns pertaining to the safety of the data. Though these systems are integrated with IoT sensors, which generally have low memory capacity, they generate humongous volumes of data per second. Thus, storing or transmitting it fast becomes challenging and critical in this regard. This way, there would be no need for centralized data storage with FL as data would be held redundantly across several clients. Hence, local processing on each client maintains the security of individual datasets. AI is considered as a robust enabler in multiple areas of applications and often acts as a communication bridge for real-world applications.

On the other hand, *blockchain* is a distributed ledger that provides security for soft components. It offers a temporary ledger that efficiently handles parameters of security and confidentiality. AI applications require a secure platform to work effectively within their application environments, and blockchain fits well into AI with the capacity of enhancing the performance of intended systems. Together, AI and blockchain produce an environment that is highly secure, through which AI agents will communicate and cooperate appropriately. There is exceptional reliability coupled with protection in delivering services not only to the existing users but also to future ones.

A core aspect of *Big Data* is in its basic characteristics: volume, velocity, and variety. One challenge in the detailed analysis of data is collecting rich data from various sources to form a total set of information. Advanced wireless networking technologies, such as 5G and 6G, which are emerging, as well as IoT and Industry 4.0, enable industries to collect data in real time as a heterogeneous source that produces abundant data. However, such a method faces several issues as data is easily tampered with during transmission, thus compromising the privacy of users. FL solves such problems by training ML models locally on data immediately available from the environment. This provides continuous training using source data without violating rigorous privacy requirements and eliminates the transfer of sensitive user data over different locations to ensure the protection of privacy.

Industry 5.0 Revolution is a vision to connect the public with smart machines. IoT devices make data gathering feasible even in far-flung places allowing easy communication with minimal extra efforts. However, the amount of data gathered is mostly raw and not in a format compatible with understanding by computations. In such cases, AI and ML algorithms get applied. These algorithms perform the critical computations, and subsequent analyses are executed extremely efficiently through edge computing. However, one of the greatest challenges is to ensure that the sensitive information is not vulnerable to attacks. Data transmitted from node to node is susceptible to being attacked at the data hub nodes. This is where FL with blockchain technology may prove helpful. Data is thus collected and processed at the local level in each device, and the results of processes are streamed to a central server. The initial parameters are then verified by blockchain technology before it releases these for processing.

3.6 CONCLUSION

Most recent technological advancements in the health sector have become a highly debated phenomenon across various research studies, mainly through the implementations of FL and AI in relevant networks. The present analysis deals fundamentally with the vast new information relating to issues of FL and AI. The relationship between FL and AI has become prominent lately as critical challenges have turned into real-time discussions on security, privacy, reliability, scalability, and confidentiality issues. FL is a new framework of distributed AI that gains major concerns in privacy and scalability enhancement for IoT services and applications. Therefore, this chapter presented the opportunities introduced by FL toward improving IoT networks through comprehensive surveys and in-depth discussions that rely on recent research work in this field. This chapter was motivated by the existing gap between comprehensive analysis and the utilization of FL within the context of IoT. In a bid to fill this gap, the chapter begins with some of the recent advances in both FL and IoT to describe their synergy integration.

Following this, the chapter identified the requirement for an up-to-date analysis by elaborating on how FL can be applied across the key services of IoT such as data sharing, data offloading and caching, attack detection, localization, mobile crowdsensing, and privacy and security of IoT systems. The chapter then focused on the recent advances in integrated FL–IoT applications in various important use case domains, ranging from smart healthcare to smart transportation, UAVs, smart cities, and smart industries. The detailed examination extensively culminated by summarizing and analyzing the major lessons learned from the research. Finally, the chapter concluded with key critical research challenges and their potential direction for future exploration. This will help raise wider interest in the new field of study and stimulate more work toward the full realization of FL–IoT and FL–AI.

REFERENCES

1. Shi, F., Ning, H., Huangfu, W., Zhang, F., Wei, D., Hong, T., & Daneshmand, M., Recent progress on the convergence of the internet of things and artificial intelligence. *IEEE Network*, 34(5), 8–15, 2020.
2. Lin, J., Yu, W., Zhang, N., Yang, X., Zhang, H., & Zhao, W., A survey on Internet of things: Architecture, enabling technologies, security and privacy, and applications. *IEEE Internet of Things Journal*, 4(5), 1125–1142, 2017.
3. Balraj, L., & Aruchamy P., An energy-aware software fault detection system based on hierarchical rule approach for enhancing quality of service in internet of things-enabled wireless sensor network. *Transactions on Emerging Telecommunications Technologies*, 35(4), 1–22, 2024.
4. Sun, Y., Peng, M., Zhou, Y., Huang, Y., & Mao, S., Application of machine learning in wireless networks: Key techniques and open issues. *IEEE Communications Surveys & Tutorials*, 21(4), 3072–3108, 2019.
5. Bhuvaneshwari, B., Balusamy, B., Dhanaraj, R. K., & Ravi, V. Artificial intelligence enabled Luong attention and Hosmer Lemeshow regression window-based attack detection in 6G. *International Journal of Communication Systems*, 36(15), 1–18, 2023.
6. Li, H., Ota, K., & Dong, M., Learning IoT in edge: Deep learning for the Internet of Things with edge computing. *IEEE Network*, 32(1), 96–101, 2018.
7. Prajith, P., Yokesh, V., Artificial intelligence-enabled fully echoed Q-routing and adaptive directional medium access control protocol for flying ad-hoc networks, *International Journal of Communication Systems*, 38, 1–18, 2025.
8. Pham, Q. V., Fang, F., Ha, V. N., Piran, M. J., Le, M., Le, L. B., Hwang, W. J., & Ding, Z., A survey of multi-access edge computing in 5G and beyond: Fundamentals, technology integration, and state-of-the-art. *IEEE Access*, 8, 116974–117017, 2020.
9. Yange, C., Lei, L., & Yuan, P., A privacy-preserving federated learning framework with lightweight and fair in IoT, *IEEE Transactions on Network and Service Management*, 21, 5843–5858, 2024.
10. Aruchamy, P., Lavina B., & Devi, S. An energy-aware link fault detection and recovery scheme for QoS enhancement in internet of things-enabled wireless sensor network, *Computers and Electrical Engineering*, 123, 1–17, 2025.
11. Li, Q., Wen, Z., Wu, Z., Hu, S., Wang, N., Li, Y., Liu, X., & He, B., A survey on federated learning systems: Vision, hype and reality for data privacy and protection. *IEEE Transactions on Knowledge and Data Engineering*, 35(4), 3347–3366, 2021.
12. Yang, Q., Liu, Y., Chen, T., & Tong, Y., Federated machine learning: Concept and applications. *ACM Transactions on Intelligent Systems and Technology (TIST)*, 10(2), 1–19, 2019.
13. Park, J., Samarakoon, S., Elgabli, A., Kim, J., Bennis, M., Kim, S. L., & Debbah, M., Communication-efficient and distributed learning over wireless networks: Principles and applications. *arXiv preprint arXiv:2008.02608*, 2020.
14. Aledhari, M., Razzak, R., Parizi, R. M., & Saeed, F., Federated learning: A survey on enabling technologies, protocols, and applications. *IEEE Access*, 8, 140699–140725, 2020.
15. Lo, S. K., Lu, Q., Wang, C., Paik, H. Y., & Zhu, L., A systematic literature review on federated machine learning: From a software engineering perspective. *ACM Computing Surveys (CSUR)*, 54(5), 1–39, 2021.
16. Lathika, V., Arunkumar, S., Balasubramani, M. Secure multi-party computation for collaborative data analysis in network security. *International Conference on Intelligent*

and Innovative Technologies in Computing, Electrical and Electronics IITCEE, 1–5, 2024. doi 10.1109/IITCEE59897.2024.10467913.

17. Lim, W. Y. B., Luong, N. C., Hoang, D. T., Jiao, Y., Liang, Y. C., Yang, Q., Niyato, D., & Miao, C., Federated learning in mobile edge networks: A comprehensive survey. *IEEE Communications Surveys & Tutorials*, 22(3), 2031–2063, 2020.
18. Zhao, Z., Feng, C., Yang, H. H., & Luo, X., Federated-learning-enabled intelligent fog radio access networks: Fundamental theory, key techniques, and future trends. *IEEE Wireless Communications*, 27(2), 22–28, 2020.
19. Niknam, S., Dhillon, H. S., & Reed, J. H., Federated learning for wireless communications: Motivation, opportunities, and challenges. *IEEE Communications Magazine*, 58(6), 46–51, 2020.
20. Mariya, O., Mariyam, O., Prasanth, A., Blockchain technology-based biometric system. *Artificial Intelligence for Blockchain and Cybersecurity Powered IoT Applications – CRC Press*, 49–168, 2024. doi 10.1201/9781003497585.
21. Nigam, C., Kirubasri, G., & Suganthi, D. Real-time object detection on edge devices using mobile neural networks. 2024 *International Conference on Intelligent and Innovative Technologies in Computing, Electrical and Electronics IITCEE*, 13, 1–4, 2024.
22. Xu, J., Glicksberg, B. S., Su, C., Walker, P., Bian, J., & Wang, F., Federated learning for healthcare informatics. *Journal of healthcare informatics research*, 5, 1–19, 2021.
23. Prasanth, A., Lakshmi, D., Dhanaraj, R. K., & Balusamy, B., Technological Advancement in Internet of Medical Things and Blockchain for Personalized Healthcare: Applications and Use Cases. *CRC Press*, 1–216, 2024.
24. Brik, B., Ksentini, A., & Bouaziz, M., Federated learning for UAVs-enabled wireless networks: Use cases, challenges, and open problems. *IEEE Access*, 8, 53841–53849, 2020.
25. McMahan, B., Moore, E., Ramage, D., Hampson, S., & y Arcas, B. A., Communication-efficient learning of deep networks from decentralized data. *Artificial Intelligence and Statistics*, 54, 1273–1282, 2017.

4 Federated Learning for Sustainable Development Using IoT/Edge Computing System

M. Yogeshwari, S. Lavanya, S. Deepa, and Afizan Bin Azman

4.1 INTRODUCTION

Smart computing plays a significant role in renovating traditional technologies into future-generation technologies with various combinations of paradigms. Among that, internet of things (IoT) provides a smart environment by connecting network devices for collecting information from the environment, analyzing the collected information, and taking necessary action for the benefit of the users. With the use of IoT and other related technologies the environment becomes smart, providing benefits to the society while improving the quality of life [1].

Edge computing aims to provide processing at the network edge, near the data source, delivering services to the users with minimum latency. But edge is resource constrained, which has limited computing capability. The review of literature in this chapter is focused on the smart environment, edge computing, which is necessary to understand the IoT device integration with edge devices. The scheduling technique has been implemented to reduce the latency of the IoT process in edge computing and is reviewed to identify the limitations in the existing techniques. The contexts in the smart environment are reviewed finally which is necessary to differentiate the IoT processes in the smart environment.

4.2 SMART ENVIRONMENT

IoT is the essential supporting element for various smart environments such as smart home, smart industry, smart city, smart hospital, smart agriculture, smart environment, tourism, safety, waste and water management, infrastructure management for smart grids and smart transportation, banking, education, and social media [2]. But they address various performance trade-offs and heterogeneous features in the IoT emerging smart environment at different pace. The applications such as smart homes could approve new technologies sooner but certain environments such as smart factories and smart hospitals could not adapt easily, since such environments face legal issues.

DOI: 10.1201/9781003532323-4

4.2.1 Components in IoT

In general, IoT is defined as things which are uniquely identifiable or addressable, which have the ability to send and receive data via internet based on standard communication protocols to perform actuation on things without intervention of humans. The basic components which are involved in realization of IoT are the things, communication technologies, and the computational element. The functionality of the IoT can be viewed as a cycle of acquiring the information from the smart environment and passing it to the upper layer for processing, and the results are actuated on the smart environment.

The sensors sense the environment using the local communication technology and the sensed data is sent to the gateway. The gateway receives the sensed data and utilizes global communication technology to send the data to a centralized remote cloud for processing.

- In a remote cloud, the raw data is processed to make useful data and it is stored in a cloud server.
- The result is sent back to the IoT device in the smart environment, performing actuations or command or control or creating triggers. The triggering mechanism can be provided by giving notification to the user by sound alerts, SMS, and email. The details of the IoT components are described in the further sections.

4.2.2 Sensors

The IoT can sense the smart environment and collect data continuously through some sensor devices which are placed at point of interest. The sensing devices can be bar code readers, RFID tags, GPS, surveillance cameras, or video sensors, that is, IP cameras, wearable sensing devices, and sensors. The sensors are chosen depending upon the requirement of the application. For example, the authors have used smoke sensors and gas sensors to detect carbon and carbon dioxide levels to detect fire occurrence. The sensors can be classified into three categories as physical sensors, virtual sensors, and logical sensors and their characteristics are demonstrated in Table 4.1.

4.2.3 IoT Device

The intelligence level comes from (i) smart devices or (ii) network communication, or (iii) smart device with through network communication. In Reference [3], the smart device must encompass the three features such as context awareness and perform autonomous computing in device connectivity either through wired or wireless networks.

IoT devices are small devices such as smart appliances, wearable devices, handheld devices, and sensors which are connected to the network. Handheld devices are smartphones, PDAs, and tablets. Smart appliances are smart TVs, smart washing

TABLE 4.1
Sensors Classifications and Its Characteristics

Characteristics	Physical	Virtual	Logical
Data source	Smart environment	Data obtained from software applications and services	Smart environment and other sensor data sources
Physical property	Tangible	Intangible	Combination of tangible and intangible sensors
Context information	Low level	High level	High level
Error level	High but error, missing value identification is easy	Low	Low
Tools	Sensor data	Advanced processing methods, email, chats, Twitter status, date and time from calendar	Sensor and advanced processing methods
Use cases	Indoor room temperature	Distance between two locations is calculated by GPS sensors	Dedicated web services to provide weather information based on physical sensors and historical data

machines, thermostats, fridge, smart fridge, etc. [4]. Wearable devices comprise devices such as smart watches, glasses, clothes, etc.

4.2.4 Actuators

An actuator is a network-connected electronic device that has capability to perform actions on the device based on the decision taken by the user-specified algorithm or rules. Actuators are used to employ physical actions such as move, open, close, carry, focus, and target [5]. For example, turn on fire alarm when smoke level is above the specified threshold. Depending on the context, the output warning can be assigned to another device instead of actuating the actuator in the IoT device. For example, notification can be given to the user with a light instead of giving a sound message in a noisy location [6].

4.2.5 Communication Technology

Ideally, sensors should communicate to the processing environment through their available input/output hardware interconnections or lightweight wireless connection. Communication technology is the fundamental element required to recognize IoT

TABLE 4.2
Communication Protocols

Range	Range	Frequency Band	Protocols
Short-range wireless	10–100 m	2.45 GHz	Zigbee, ZWave, Ant+, WirelessHart, BLE,NFC, RFID, EnOcean, Insteon, MiWi, DigiMesh, Thread, 6LowPAN, IEEE 802.15.4, ITU-T G.9959
Short-/medium-range wireless	100–1000 m	2.4–900 GHz	HaLow, LTE-Advanced, 802.11/a/b/g/n/ac, 802.11af, ah(TVWS), IEEE 802.11ah (60 Hz), WiFi, DECT-ULE
Medium range	Approximately 5–15 km	NA	Wi-SUN, ZigbeeNan, NB-IoT, LoRaWAN, DASH7 Symphony Link
Long-range wireless	Up to 100 km	868 and 915 MHz	Cellular 2G/3G/4G/5G LPWAN, VSAT, Sigfox, LoRa, IEEE802.15.4, Bluetooth/LE

services. Sensing environment has various characteristics and it generates different forms of data, so it is not practical to apply a standard communication technology for all the IoT applications [7]. Typically, the IoT devices should operate using low power in the presence of lossy and noisy communication links [8–11]. Certain key aspects must be considered in network connectivity such as range, data rate, frequency, security, and power.

Table 4.2 shows the examples of communication protocols used for the IoT and they are classified as:

1. Short-range or wireless personal area network
2. Long-range or wireless wide area network

Short range wireless protocols are used to communicate between IoT devices and edge, whereas long- and medium-range protocols can be used between edge devices and to the outside world environment. The range of wired communication depends upon the distance of wired connection [12].

4.2.6 Computational Unit

The IoT device generates a huge amount of data which needs to be stored and processed for smart applications, intelligent decision-making, and for actuation [13]. Applications like smart grid and real-time monitoring as well as video gaming needs huge data storage, high bandwidth for data flow, and real time data analysis. Some IoT applications need intelligent algorithms to make decisions to provide service to users. The storage for huge data from different heterogeneous devices and real-time computing by intelligent analytic algorithms is provided by cloud computing paradigm [14]. The data collected from IoT devices are gathered, stored, and analyzed in the cloud with large computing devices to provide useful information. Cloud provides relevant information to the users by actuating actuators in the IoT devices [15].

4.3 EVOLUTION OF EDGE COMPUTING

The data from the smart environment must be processed in real time to achieve reliable and qualitative service. The huge amount of IoT data generated does not provide instantaneous real-time service and diminishes the quality of service (QoS). This challenge faced by execution of IoT process in traditional systems can be superseded by the advanced technology called edge computing. Edge computing enables computers to compute and store data near the data source at the network edge and provide real-time service. Many researchers have contributed their concepts based on their needs [16].

Initially, the content delivery network (CDN) used cache servers to deliver the web documents to the nearby users by placing geographically distributed servers to deliver services more quickly by reducing network congestion [17]. In 2000, peer-to-peer network overlays were proposed to process decentralized distributed applications without centralized control. It uses a distributed hash control mechanism to identify the peers in the peer-to-peer network. Each peer maintains neighbor peer node ID and IP addresses enabling fault-tolerant routing and object location. These systems use the network proximity connections and therefore use the neighbor peers, aiming to reduce the latency of the applications [18]. But commercial technologies based on peer-to-peer networks were not recognized because of its illegitimate file sharing.

Cloud computing is introduced to deploy resources such as infrastructure, application, and data in a centralized environment via the internet and they are provided to users based on their demand with the help of service providers. Cloud computing introduced new parameters such as virtualization, load balancing, and storage capacity [19]. The cloud uses virtualization mechanisms to increase the efficiency of cloud servers, limit the use of resources, and administrative costs. Since it is centralized, the latency for application is more because of network congestion and therefore it does not support instantaneous decision-making. The cloudlet can be accessed by mobile devices to provide services with less latency and high bandwidth.

The cloudlets are placed in the public places and offer services immediately using virtual machine technologies [20]. The network latency can be reduced, but mobility remains the major challenge in cloudlet. Since the mobile edge base stations are placed in a wireless access network and in the proximity of mobile users, the services can be provided to the users with less latency and high bandwidth and aims to improve the quality of service and user experience.

Multi-access edge computing (MEC) is similar to edge computing in such a way that the mobile devices which are acting as edge servers are placed between cloud and edge which aims to reduce the communication links between cloud servers and mobile users. But the major challenge is the energy required by the mobile device to offer services. Another challenge is the number of mobile edge devices required for performing large computational tasks and its performance issues.

Fog nodes are placed in between traditional cloud servers and the smart devices and thus act as an intermediate layer providing computing power at the local area network where the data are processed and the results are transmitted to appropriate devices. Edge computing was the successor of fog computing, and the authors argue that edge computing cannot be substituted by fog computing. In case of edge computing, computation is envisioned for the things, and fog computing is intended for the infrastructure side. The basic foundation is computation takes place at the proximity of sensors. The difference between these two computing lies in where the computation is done.

In fog computing, the data is collected and transferred to the LAN where the computation occurs. The edge computing processing occurs at the edge of the network, which is very close to the data source from where the data is generated. In edge computing, the computation takes place in between the location where data is transferred to the LAN, probably at the hub or gateway or microcontroller or automation controller and they are referred as edge devices. The processing capacity and storage capacity of the edge is smaller than the fog. The fog node capacity is less than the cloud.

4.3.1 Edge Computing Definition

Edge computing allows the processing of the IoT data at the edge of the network, with integration of cloud infrastructure. Various authors have defined edge definition in different ways. Edge computing is a computing paradigm, intending to perform computation near the IoT device where data evolves, enabling less latency and providing high bandwidth, reducing communication cost while ensuring security and privacy to the end-user handling the IoT device and the cloud.

4.3.2 Edge Devices

Edge computing is performed with edge devices. Edge device is resource-constrained and has limited computing capability. Individual human-controlled devices such as tablets, desktop PCs, nano data centers, laptops and smartphones, stable computing devices such as routers and media centers are incorporated. Any device which has computational power and network properties in between IoT device and remote cloud and computational device must be in the vicinity of the data source. For example, in the case of smart health to monitor the status of a patient, a smart phone can be taken as an edge device; and in case of smart home, a gateway can be taken as an edge device.

Edge devices can be any nodes between IoT device and cloud such as routers, base stations, switches, and access points and traffic aggregation points. Node powers computing resources on end devices such as smartphones, Google glass, and home storage devices. A dedicated edge server is attached to the access point or base station. An edge is any hub with limited computing capability which is linked to the smart environment referred to as edge device.

The edge device has limited computing capability, all the processes from the smart environment cannot be processed at edge. So, it is necessary to review the various processes generated from the smart environment and among all the processes which process can be scheduled at edge.

4.4 VARIOUS PROCESSES IN SMART ENVIRONMENT

More research has been done on various types of IoT processes in the smart environment. This section discusses the process classification with different parameters. The processes in the smart environment are identified and listed in Tables 4.3 and 4.4. The authors classified the process into light-weight and

TABLE 4.3
Classification of IoT Process with the Parameters

Process Classification	Parameters
High, medium, and low process	Context
Light- and heavy-weight process	Processing delay
Normal and emergency process	Traffic
Four classes	Mission-critical and delay-sensitive process
Control application, monitoring application, and enquiry application	Type of information managed by the applications, recipient either person or resource, criticality

TABLE 4.4
IoT Processes in Smart Environment

Applications	Mission-Critical Process	Delay-Sensitive Process	Delay-Tolerant Process
Smart home	Fire and smoke detection	Smart fan, air cooler, light, kettle, water tank controller	Smart power meter
Smart dairy farming	Disease detection	Virtual fence management, feed management	Monitor weather changes and grass growth
Military health services in battlefield	Detection of isolated soldiers, soldiers' health condition	Weapon status, geographical location of soldier	Not mentioned any process
Smart hospital	Detection of emergency patients	Not mentioned any process	Sending health reports to the database
Atomic power station	Detection of radioactive discharges	Detection of trespassers	Not mentioned any process
Smart ATM	Smoke and fire detection, intrusion detection	Battery health check-up, diagnostics of ATM system	Smart meter

heavy-weight processing depending upon the processing delay latency required by the IoT process.

The IoT process is differentiated based on whether the process is mission critical or not and the delay required by the IoT process. The process is classified in the IoT environment depending upon the latency and the spatial location. The IoT applications can be classified into different methods, and control application is considered as a mission-critical process which is highly critical and needs emergency management with quick response time and less fault rate.

High priority must be given to the mission-critical IoT process, since it can include human, economic, and environmental loss. The mission-critical process depends upon the application, it may affect human lives or it may cause or damage any system or result in failure of project or business. While looking at the mission-critical application, if it fails it causes catastrophic damages, affecting human life, dealing with safety of human life and any system or project or organization.

4.5 PROCESS SCHEDULING

Research on process scheduling in edge computing is still at a premature stage. Process scheduling in edge computing plays a main role to decide whether the process is executed in edge or cloud server, focusing on reducing the latency of the IoT process generated from the smart environment. The process scheduling on the edge can be classified into heuristics algorithm and meta-heuristic algorithms.

The process can be scheduled easily by using heuristics and the solution provided by the algorithm is best. The meta-heuristics algorithms are enriched algorithms and they are based on heuristic algorithm. These algorithms are used when the information is imperfect and based on few assumptions. These algorithms have high complexity when compared with heuristics algorithms.

In edge computing, the researchers have used the existing traditional scheduling algorithms such as FCFS, SJF and round robin algorithms, min-min, max-min. So the literature study concentrates mainly on various scheduling algorithms for reducing the latency of the IoT process generated in the smart environment.

4.5.1 Scheduling Algorithms in Edge

In the IoT-based smart environment, things such as sensors, smart phones, smart appliances, smart devices, connected cars, and any other embedded devices transfer information to high-performance computing capability centers, generating data on an enormous scale. Various surveys infer and express that in the future (2050) around 100 billion IoT devices will host many applications in everyday smart devices. This delay can be reduced by edge computing in which IoT processes are executed at edge. This section describes the various scheduling algorithms at edge in a smart environment with scheduling parameters and limitations.

4.5.1.1 Priority-Based Scheduling

This algorithm is used by priority-based scheduling among two applications. The method applied is application aware, in which priority is set among two applications.

High and low priorities are set in two applications, so when the data packets arrive at the edge, the high-priority application is processed at edge and low-priority application is sent to the cloud for processing. The technique proposed is pre-defining the priority at application level without considering any other properties of the applications such as security, reliability, sensed environment values. This algorithm is applied for smart hospitals which focus on giving high priority to patients based on the sensors used by them. Based on the sensor values from the patients, the process is classified as an emergency and non-emergency process.

(a) Scheduling Policy

The latency of a process can be reduced by selecting edge node which will result in providing least latency between source node and edge node. The author has defined a mathematical model to verify whether the edge architecture provides less latency for the user request. They proved that communication delay can be reduced by applying least latency policy between source and edge node. It has proposed to schedule processes to the edge node with shortest response time and shortest latency in between edge nodes. Among the two algorithms, the shortest response time provides less latency in executing the process at the edge.

(b) Software-Defined Networking (SDN)

Software-defined networking (SDN) architecture helps to acquire the periodic information about edge server and user entity processing power, queueing condition and network condition. This helps to minimize the latency of processing applications by finding the nearby server. The user-requested processes are scheduled among edge servers using SDN. But this work focuses on general scheduling and thus not applied in scheduling in a smart environment.

(c) Context-Based Approach

Application's context-sharing mechanism works by means of finding suitable nearby edge nodes having the same type of application context instance and then executing the process within the defined deadline to minimize the latency of applications. But the context required for the IoT process in smart environment is not mentioned. Thus, context-based scheduling helps to minimize the latency of a process by gathering context information about the environment variable in focus [21].

The context parameters taken are number of replicas of data, remaining storage capacity of the system, current fluctuation, and the application. Here data placement plays the major role for scheduling the process. Data is placed in duplication, prior to process scheduling in the edge systems. Replication of data in edge systems is an overhead in this scheduling algorithm. So, the authors have not focused on context attributes to determine criticality of a process.

(d) Content Delivery Network (CDN)

Content delivery network (CDN) is a distributed web cache server which provides less latency than cloud, but they are used to distribute static web contents. The scheduled video streams at the edge reduce the latency of applications. They use edge node and

CDN node to compare the efficiency of the algorithm. Their results prove that latency of using the edge is less than using CDN nodes.

(e) Fuzzy Logic Technique

Fuzziness is dealt with uncertainty in the IoT environment, since the environment varies dynamically and the attribute values in this environment vary continuously. The fuzzy logic approach helps to think like a human and have a decision-making mechanism to provide resultant value where the environment is vague or uncertain [22]. This study applied fuzzy approach to classify the process by setting user-defined priority to the process as one of the fuzzy attributes and scheduling the process on the edge and cloud nodes. The authors do not focus on the attributes for identifying the priority of user-defined process for the IoT process.

Two-stage fuzzy mechanisms are applied for management of resources at the edge and cloud. But the authors have not focused on a smart environment to identify the criticality level of the process, they applied the fuzzy logic approach to manage the edge and cloud resources. The fuzzy approach is used to assign priority to edges for the execution of IoT processes generated in IoT based on traffic available in the network edge, not based on the criticality of the process

Smart environment comprises various IoT processes, but the criticality of the IoT process is based on only the sensor value in an IoT device in the existing scenario. But few researches have been done to find the criticality of an IoT process based on sensor value and delay tolerance values. Therefore, there is a need to find the criticality of the process in the current scenario to schedule the IoT process at edge. To develop a smart scheduling algorithm for smart environment, an intelligent approach to find the criticality of the IoT process with context awareness must be applied. Therefore, a detailed study is made on context attributes for finding the criticality of the process in a smart environment.

4.5.2 Study of Existing Context Elements

The study of context awareness was initiated with the help of context-aware applications, which are used to find the location of a user to forward the phone calls to nearby users. Many researchers then focused on context awareness in their work to apply in different applications with different scenarios. Later it was used in pervasive computing to apply in dynamic environments, where the user requirements were changing. Therefore, based on studies made from literature, context can be classified as:

1. Primary contexts
2. Context characteristics
3. Contexts source

4.5.2.1 Primary Contexts

In a real-time environment, the basic context elements are derived from various sources, and as the environment evolves dynamically, these elements continue to grow and expand. The dynamically changing environment predicts the context elements required for a particular application. Location context is used to find the location of a user or it can be a room or any building. The word context awareness is used to describe the context by means of location. The location of the user is derived by using GPS coordinates (ATM center), specific address (smart home of a user), user favorites (users' bank branch, user service center), common names (university, college, school, hospital).

Things context as resources available near the use location such as smart appliances in smart homes and things in smart shops, which illustrate the things context as any physical component such as computer, laptop, tablet, and mobile. Nearby things are defined as things context in the real-time environment:

1. **Critical states**: Critical states refer to trigger an alert if a condition mentioned in an environment is satisfied.
2. **Spirits**: Spirits refer to user-stated context or pre-defined commands, if the context cannot be detected by the computational device.
3. **Historical context**: Historical context refers to store context information. Historical context information helps to find the activities, favorites, desires of the user, patterns, and trends in smart environment.
4. **Activity**: The context represents the essential characteristics of an entity such as work done by a device, service delivered by an entity. The entity can be people, place, or thing.

4.5.2.2 Context Characteristics

Context information mostly is based on homogeneous information such as location and sensed data. The context type for pervasive computing systems are categorized as sensed, static, profiled, and derived on the basis of the characteristics of the context.

Sensed

The data collected using sensors, such as motion detection using PIR sensors, is referred to as sensed context. The values from the sensor will be changing in a dynamic environment.

Static

The information which does not change over time, such as sensor measurement range, manufacturer name, sensor configuration.

Profiled: The information which changes over time with low frequency or once per month such as transaction per month in an ATM, energy consumption in smart home.

Derived: The data derived from basic context such as location of a user, using user mobile or GPS coordinates.

Contexts Dimension

The context in other dimensions is classified as external and internal context.

External or physical context: The data obtained from physical sensors such as temperature, pressure, location, sound is referred to as physical or external context.

Internal or logical context: The information listed by the user or information obtained by interacting with the user such as user emotional state, goals, business process are referred to as internal context. They are gathered from user documents and readouts from webpages.

Mostly external context is used in context-aware systems, since they are obtained in real time and they provide useful information.

IoT devices consist of sensors, smart appliances, and actuators and their communication protocols are heterogeneous and ubiquitous which are multifaceted and interpretation of these components and data generated are difficult to understand. Therefore, literature review is made to understand the contexts in IoT-based smart environment.

4.5.2.3 Context Awareness in IoT

IoT makes the environment smart by providing services to the people by connecting things through various communication protocols. The smart environment is embedded with intelligence and is explored to deliver contextualized user experience. Thus, the smart environment encompasses intelligence and context awareness. Context awareness in smart environment refers to collection of context data from the sensors in the IoT device to provide services to the users.

Contexts in IoT

Context awareness plays a major role in developing the smart environment and its associated applications. Context-aware computing helps to understand sensor data and system behavior. The existing context and the new context used in IoT are enumerated in Table 4.5.

TABLE 4.5
Contexts in IoT-Based Smart Environment

Application	Existing Context Elements	New Context in IoT
Smart home	Sensor data	Human behavior, user's facial expression, user preference
Smart home	Sensor data, time context	No complex data
Smart home	User, location, time	Service context
Task placement based on data replication	Edge device	Data size, Number of replicate data, current fluctuation, battery status
AAL	Sensor data, time, location	Camera images
Cost estimation	Location, wearable devices, sensor data	Application, user preferences
Smart city	Number of sensors, sensor data location	Sensor characteristics: accuracy, reliability, cost

The research work carried out in context-based IoT is explained in this section. Various contexts in road-side units such as smart traffic light, traffic panels, video cameras, vehicle speed, consumption, user's activity, current time, weather, user's location, user's schedule and preferences, and service context are applied at three levels in smart city application. Context based on number of data replication in distributed edge such as data size, remaining storage capacity, current fluctuation, and application is defined.

Cost estimation is done for computing devices and derived various contexts such as sensor data to monitor vital signs, IoT device for battery status, environmental information such as temperature, KPI for user preferences, IoT network level for local- or global-scale policy requirements, and proximity detection. Additionally, other contexts are also used such as location, health conditions, connected peripherals, available communication links, and application contexts for health monitoring. Context based on sensor characteristics, such as accuracy, reliability, cost, is also defined.

4.6 CONCLUSION

The functionality of the smart environment has been discussed in detail with the components of the IoT. The evolution of edge computing and the various edge devices are presented in brief to understand the edge computing capability. In order to understand the various processes in the smart environment, smart applications with their process were reviewed in brief. Various scheduling algorithms that are focused mainly to reduce the latency of the IoT process at the edge were discussed. The contexts required for the smart environment have been presented. Finally, the chapter explored the research work that was carried out in context-based IoT applications.

REFERENCES

1. Palagan, C., Antony, S., Jereesha, S., Predictive analysis-based sustainable waste management in smart cities using IoT edge computing and blockchain technology, *Computers in Industry*, 166, 1–19, 2025.
2. Aruchamy, P., Lavina, B., Devi, S. An energy-aware link fault detection and recovery scheme for QoS enhancement in internet of things-enabled wireless sensor network, *Computers and Electrical Engineering*, 123, 1–17, 2025.
3. Pererab, C., Chi, H., Jayawardena, S. A survey on internet of things from industrial market perspective, *IEEE Access*, 3, 1–18, 2015.
4. Guillermo, S., Juan, P., Arturo R., Evolution, detection and analysis of malware for smart devices, *IEEE Communications Surveys & Tutorials*, 16, 1–19, 2013.
5. García, G., Daniel, M., Cristina, P. A review about smart objects, sensors, and actuators, *International Journal of Interactive Multimedia and Artificial Intelligence*, 4(3), 1–19, 2017.
6. Sasikala, P., Aravinth, T. S., Dharani, A. IoT-based Smart Intelligent Vehicle Systems, *2020 3rd International Conference on Intelligent Sustainable Systems (ICISS)*, Thoothukudi, India, 2020, 1094–1097.

7. Mudassar, S., Muhammad, U., Zafar, A. Design patterns and general video game level generation, *International Journal of Advanced Computer Science and Applications*, 8(9), 1–19, 2017. https://doi.org/10.14569/IJACSA.2017.080952.
8. Sinha, R.D., Kumar, B., Hemanth Kumar, K. A context-aware fog enabled scheme for real-time cross-vertical IoT applications, *IEEE IoT Journal*, 6(2), 3371–3381, 2019.
9. Ghanbari, Z., Navimipour, N., Hosseinzadeh, M., A new lightweight routing protocol for internet of mobile things based on low power and lossy network using a fuzzy-logic method, *Pervasive and Mobile Computing*, 97, 1–17, 2024.
10. Jeyaselvi, M., Dhanaraj, R. K., Sathya, M., Memon, F. H., Krishnasamy, L., A highly secured intrusion detection system for IoT using EXPSO-STFA feature selection for LAANN to detect attacks, *Cluster Computing*, 26(1), 559–574, 2022.
11. Mocnej, J. Impact of edge computing paradigm on energy consumption in IoT, *IFAC-PapersOnLine*, 51(6), 162–167, 2018.
12. Perera, C., Harold, L., Jayawardena, M. A survey on internet of things from industrial market perspective, *IEEE Access*, 2, 1660–1679, 2014.
13. Jayavardhana, G., Rajkumar, B. Internet of things (IoT): A vision, architectural elements, and future directions, *Future Generation Computer Systems*, 29(7), 1645–1660, 2013.
14. Kyoochun, L. The internet of things (IoT): Applications, investments, and challenges for enterprises, *Business Horizons*, 58(4), 431–440, 2015.
15. Prasanth, A., Jayachitra, S. A novel multi-objective optimization strategy for enhancing quality of service in IoT enabled WSN applications, *Peer-to-Peer Networking and Applications*, 13, 1905–1920, 2020.
16. Abdul-Qawy, A., and Potluri, A. Performance analysis of parameters affecting power efficiency in networks, *2015 7th International Conference on Communication Systems and Networks (COMSNETS)*, Bangalore, India, pp. 1–8, 2015.
17. Sinha Roy, D., Behera, R. K., Reddy, K. H., and Buyya, R. A context-aware fog enabled scheme for real-time cross-vertical IoT applications. *IEEE Internet of Things Journal*, 6, 2400–2412, 2019.
18. Nigam, C., Kirubasri, G., Aeron, A., & Suganthi, D. Real-time object detection on edge devices using mobile neural networks. *2024 International Conference on Intelligent and Innovative Technologies in Computing, Electrical and Electronics (IITCEE)*, 13, 1–4, 2024.
19. Nikshya, J., Sanmugavalli, P., Hussain, A. An hybrid combination of ontology along with RDF technology for ICU systems through CBM. *2024 4th International Conference on Advance Computing and Innovative Technologies in Engineering (ICACITE)*, 2, 151–154, 2024.
20. Banu, F., Udhayamoorthi, M., Nallagattla, V. R. G. Enhancing low latency and high performance in clustered-based wireless sensor network routing using genetic algorithm for 5G networks. *Optical and Quantum Electronics* 55, 1048, 2023.
21. García-Valls, M., Dubey, A. Introducing the new paradigm of social dispersed computing: Applications, technologies and challenges, *Journal of Systems Architecture*, 91, 83–102, 2018.
22. Jayachitra, S., Kumar, G. Seismic Image Pattern Analysis Using Fuzzy Logic Controller, *2020 3rd International Conference on Intelligent Sustainable Systems (ICISS)*, Thoothukudi, India, pp. 998–1002, 2020.

5 Advances in 5G/6G-Enabled Federated Reinforcement Learning in IoT

P. Arul, V. Yokesh, N. Sathish, and Pham Chien Thang

5.1 INTRODUCTION

The availability of 5G's and 6G's and the standout features that they present is improving the capacity of federated reinforcement learning (FRL) in Internet of Things (IoT). These are some of the activities that have been made possible by the need for high speed, low delay, and more secure connectivity of distributed devices. In this regard, federated learning (FL) solves the issues to do with privacy and security regarding IoT information through decentralized procedures [1]. Further, it may train other learning algorithms based on local information without data exchange through intelligent AI-based algorithms.

Presently, it is estimated that there are about 65 billion devices that are connected to the internet with estimations of this reaching 100 billion by the year 2030 [2]. The use of gadgets is creating numerous issues with concerns about assaults on networking systems, data breaches, complications, battery issues, as well as power issues. On the other hand, it is challenging for broad IoT devices to be connected with 5G mobile-cellular networks [3]. Therefore, it is expected that cellular 6G networks will increase the connection capacities by ten to a hundredfold relative to the 5G systems, which will help to demand a convergence of Communication, Computing as well as Caching (3C).

As the world embraces smarter and more connected technologies, IoT applications like augmented and virtual reality, autonomous vehicles, and smart devices are becoming increasingly prominent. However, the immense data needs, real-time processing demands, and low-latency requirements of these applications are proving to be too much for current 5G networks. This has created a pressing need for 6G technology, which aims to push boundaries with faster data speeds, near-instantaneous response times, and incredibly precise device positioning. These advancements will pave the way for seamless integration of IoT into our daily lives, transforming industries and redefining how we interact with technology [4].

 DOI: 10.1201/9781003532323-5

5.1.1 Overview of IoT and Its Evolution

The IoT refers to the physical objects that are connected to the internet and are capable of exchanging information. These devices, including domestic appliances and industrial sensors, are equipped with sensors, software, and other technologies for data acquisition and sharing. The progression of IoT has been outlined by timely evolutions in connectivity, compute power, and physical size. When it started as simply machine-to-machine communication or M2M the IoT has grown into an intelligent system of interconnected homes, cities, hospitals, and industries. With the emergence of IoT-connected devices and advancement in IT technologies, such as cloud computing and big data analysis, the decision-making process has gradually shifted toward data-based where and when decisions and are increasingly becoming automated and optimized.

5.1.2 Importance of IoT in the Context of 5G/6G Networks

With the rise of IoT, the need for reliable, fast, and low-latency communication networks has emerged [5]. It is where 5G and, in the future, 6G networks come into play the crucial part. 5G brings distinct advantages in the form of significantly higher data speed, lower latencies, and the ability to support a significantly large number of connected devices at the same time. These attributes are essential when enabling real-time communication and processing as evident in the case of IoT, self-driving cars, and smart power stations among other applications in the health sector.

The future 6G networks are expected to expand these limits even more with the inclusion of such features as terahertz communication and AI in networking to support even more complex IoT applications that require immediate responses to data processing. This conjoining of IoT with 5G/6G networks is expected to revolutionize companies through enhancing more efficient, flexible, and smart systems. Some of the most crucial regions of IoT, where 6G deployment is unavoidable, are depicted in Figure 5.1.

5.1.3 Introduction to Federated Learning and Reinforcement Learning

In general, FL and RL are the two modern approaches of machine learning (ML). FL is a distributed learning approach in which many edge devices contribute to building a global model without sharing local data with other devices; this makes FL especially useful in environments where data confidentiality and integrity are paramount [6]. On the other hand, RL involves the training of an agent to perform sequential actions by interacting with the environment with the purpose of maximizing cumulative reward. RL is more useful in a stochastic environment where the best actions to be taken are learnt through experience. FL and RL are seen to have great potential across different domains and when combined, there is a robust mechanism for decentralized learning in complicated distributed systems, including IoT, which is being discussed in this chapter.

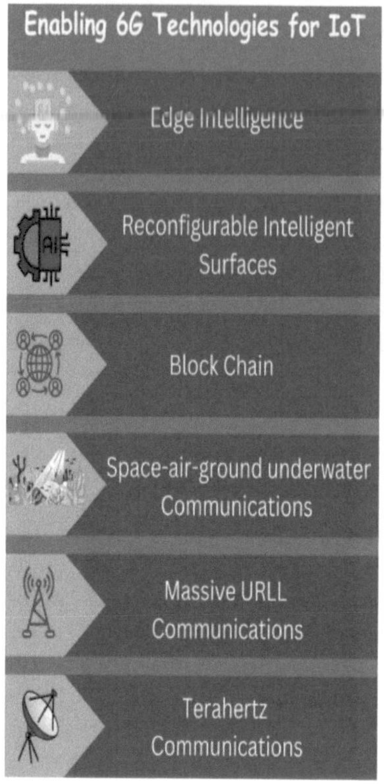

FIGURE 5.1 Fundamental 6G technologies for IoT.

5.1.4 Motivation for Integrating FL and RL in IoT Environments

The adoption of FL and RL inside IoT ecosystems is due to the various issues and opportunities the IoT networks present. By doing so, IoT devices generate vast amounts of data that are often scattered, heterogeneous, and/or contain confidential information [7]. To overcome privacy concerns and reduce the burden of communication costs, FL allows devices to collaborate to learn a common global model keeping data locally. Real-time adaptive decision-making is required in many IoT applications, such as intelligent transport systems (ITS), smart power grids, industrial IoT, and self-driven cars, hence RL comes in handy [8].

By combining the concepts of FL and RL, IoT devices are able to learn from dispersed data in a decentralized manner and adapt their behavior in reaction to real-time inputs from the environment. This methodology enables the build of intelligently scalable IoT systems that are able to manage themselves while adapting their operations for maximum efficiency and privacy/securing of the collected data. The combination of FL and RL in the IoT is considered to be a giant leap in the evolution of intelligent systems capable of performing their tasks in rather complex embedded environments in a decentralized manner.

5.2 FUNDAMENTALS OF FEDERATED LEARNING

FL enables many devices or edge nodes to train a model without sharing their local data with other nodes. FL preserves the data locally in the devices it is produced and only sends model updates containing gradients or other model parameters to a central server or aggregator as opposed to traditional centralized ML. FL mitigates data breaches and General Data Protection Regulation (GDPR) noncompliance by storing the data locally.

The various stages of the FL process are shown in Figure 5.2 and it is elaborated in the following:

1. *Initialization:* A main server creates a global model, which is deployed over the client devices.
2. *Local Training:* When working with particular clients, the local data is trained for the model, and parameters are changed depending on the dataset of the client.
3. *Model Aggregation:* In the model aggregation step, locally trained updates are sent back to the central server for integration in order to update the global model. One of the often-used processing methods is weighted averaging where updates are weighted by the size of the local dataset.
4. *Model Update:* Exchange with clients, update the global model, and report new solutions to all clients concerned until it converges.

FL is particularly suitable for distributed data source scenarios such as mobile terminals, IoT, and edge computing. It makes the development of secure and efficient big data computing possible and applicable in the healthcare, finance, and telecommunication industries. Besides privacy, FL also solves the issues of communication and computation complexity. FL does not transmit massive datasets from the participating devices to a central server, thus cutting on communication overhead, especially in settings characterized by low bandwidth and high latency [9]. FL leverages the computational ability of dispersed devices in terms of resource employment as well as training of large-scale models.

FL also involves dealing with non-independent and identically distributed (non-IID) data, guaranteeing model convergence across different devices and addressing some security concerns such as model poisoning. FL study intends to strengthen, enlarge, and secure this decentralized system to remove these challenges. FL modifies

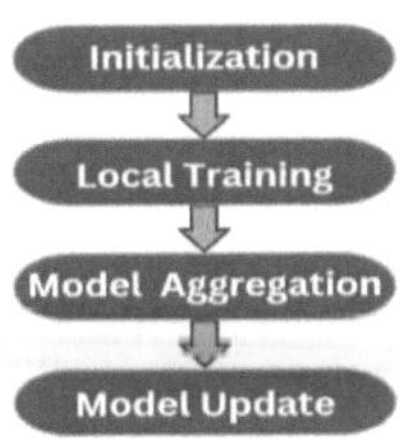

FIGURE 5.2 FL process.

the way that teachers deliver ML models to learners, and it offers a solution to data privacy and security issues and top-batch training.

5.2.1 Principles of Federated Learning

FL is categorized as a decentralized ML approach where countless clients such as mobile devices or edge nodes all learn from the same model, but no information from any client's data is shared. Unlike the former approach in which the principal sends unprocessed stochastically data back to a central server, each client independently rebroadcasts only adjusted model parameters up to the server. It aggregates these changes to improve the global model, enhance data privacy, reduce communication costs, and use datasets from different clients [10]. FL is highly beneficial in domains that are privacy-preserving in nature like healthcare and finance [11].

Basic concepts of FL include data protection and data security because data remains stored on personal devices and thus does not have a high likelihood of data breach. Decentralized learning is also the main methodology of how nodes train and update a global model at the same time. FL saves communication overhead and allows processing independently of other strata. Which is also robust and fault tolerant, it can also scale to millions of devices [12]. FL brings personalization and standardization whereby models can be trained locally while maintaining the data privacy of the data it deals with by keeping it locally.

5.2.2 Challenges and Opportunities of FL in IoT

FL restores decentralized ML in IoT paradigms and changes the approach but also has distinct challenges and prospects. The issue of managing IoT devices is challenging because they differ in processing capabilities, battery, and network. It outlines different technologies, which hinder training across devices or can lead to compromised performance and accuracy in skills. FL contains data of IoT devices; thus, FL data privacy and security are important. FL is decentralized, which eliminates some risks and adds a new level of risk such as model poisoning and communication overhead. Consequently, FL presents vast IoT opportunities even with these hindrances [13]. It enables real-time, personalized AI models that adapt to device contexts without transferring data, improving privacy. FL can also reduce central data storage, saving money and boosting data sovereignty. FL in IoT will likely become increasingly practical as 5G and edge computing technologies evolve, enabling smarter, more responsive, and privacy-preserving IoT systems.

5.2.2.1 Challenges of Federated Learning in IoT

Complex model training is challenging on many IoT devices due to their low computing power. Memory and storage limit the amount and complexity of models these devices can manage. IoT devices often operate in locations with poor network connectivity, causing network latency and central server synchronization delays [14]. Battery-powered IoT devices that communicate with the server frequently can consume energy.

IoT data heterogeneity affects FL efforts. Non-IID IoT data makes model convergence and generalization difficult. IoT networks have many devices with different sensors and data formats, making local model updates problematic. Cyber security for IoT devices has become a major concern since they are more vulnerable to model poisoning and adversarial attacks. Data is stored locally in FL, and while inference attacks may need additional privacy-preserving strategies.

Last but not least, FL frameworks with IoT devices have practical problems regarding the organization and scale of logistics. To scale learning with hundreds or millions of devices, coordination and consolidation are entailed. There are loners, devices that might have lower processing speeds or a weaker connection, which may slow down or cause inconsistency with model updates, which might be a challenge at the large-scale deployment of the coordination systems.

5.2.2.2 Opportunities of Federated Learning in IoT

FL tremendously enhances the privacy of IoT applications given that computations are done locally. Devices can feed into a global model at the same time as data is stored on the local premise thereby lowering the chance of leakage. This localized approach can make IoT systems in compliance with stringent data privacy laws that are presently in force around the world, including the GDPR, since the data is kept on the device and the user retains the authority over it. With privacy emerging as a key issue in most IoT networks, FL offers a structure that is compatible with these policies.

It also allows for the learning process to be distributed over various devices so that one serves as a mechanical extension of the other. This decentralized approach means that IoT devices are training more generalizable models together while at the same time updating them in real time [15]. The outcome is better model behavior in conditions where changes are expected to occur from time to time, useful for projects like self-driving automobiles, or intelligent homes. Furthermore, FL eliminates the need for the exchange of a massive amount of data between devices and central servers to achieve effective training, thus consuming minimum bandwidth and energy necessary in tight IoT surroundings. Realization of FL with edge computing also makes the framework even more efficient in terms of computation offloading to more capable edge servers reducing latency.

Furthermore, FL is enhanced to build personal and contextual models that provide IoT devices with the probability to realize global models with respect to the context and user behavior. As a result, there is increased satisfaction by the users in areas such as healthcare, smart cities, and energy, among other areas where tailored solutions are needed. Over time, FL's constituent technologies are changing in response to these challenges as new security technologies for distributed learning, including encryption and differential privacy, continue to build increased robustness into IoT networks against adversarial attacks and model poisoning. These innovations will allow FL to enable safe and efficient AI solutions in various industries, including smart transportation in urban environments and remote healthcare for patients in smart health-based IoT applications. As has been seen, there are still issues to be addressed; however,

FL has the potential to change the way that IoT ecosystems can be architected and deployed.

5.2.3 FL Applications in IoT

FL's ability to enable ML in distributed devices and respect data privacy, FL is crucial in the context of IoT. Here are some FL applications and use cases in IoT.

5.2.3.1 Smart City and Home Apps

Energy Management: Smart meters and thermostats can use FL to optimize energy consumption in smart homes and cities without revealing personal data with central servers. Devices can collaborate to estimate peak usage and modify settings to save energy. *Traffic Management*: Smart city IoT sensors can analyze traffic trends and predict congestion in real time using FL [16]. This information can optimize traffic light timings and route vehicles more efficiently while localizing sensor data.

5.2.3.2 Health Care/Wearables

Personalized Health Monitoring: FL can monitor and analyze users' health indicators (e.g., heart rate, sleep habits) without sharing sensitive health data on wearables. The devices can develop health monitoring algorithms together and personalize them for each user. In healthcare, IoT devices can remotely monitor patients and submit model updates to a central server for worldwide modeling. This enhances diagnosis and prescription of individualized therapeutic plans while at the same time shoring up patient information.

5.2.3.3 Industrial IoT (IIoT) Predictive Maintenance

FL processes chaotically generated machine data to detect impending equipment failure in industrial settings. Locally, the problem is predicted on each machine and the global predictive maintenance model is refined. It provides increased time efficiency and a decrease in the need for maintenance while never sharing internal details between the devices. *Supply Chain Optimization*: Taking advantage of localized data at the warehouses and the trucks, FL can enhance the supply of IoT devices in a chain by improving the logistics and inventory of trucks. This has the added advantage of simplifying operations while ensuring that all stakeholder information is kept secure.

5.2.3.4 Autonomous Cars

Self-Driving Car Collaborative Learning: FL allows self-driving cars to understand data of many cars driving without sharing trip information. It also enhances functions such as objects' recognition and the vehicle's trajectory computation while preserving sensitive geographical information. *Vehicle-to-Infrastructure (V2I) Communication*: Traffic lights, road sensors, and even cars can help to train models to enhance the flows on roads and to increase protection [17]. FL allows this without aggregating these devices' large datasets at the central FC layer.

5.2.3.5 Agriculture

In agriculture, specific gadgets, such as the soil sensors and drones, can utilize FL in the enhancement of crop husbandry through examination of the local environmental parameters. Crop yield forecasts and irrigation schedules help farmers without sharing valuable agricultural information. FL may use pest detection algorithms trained from data from various farms to more accurately predict pest infestation while at the same time securing agriculture data.

5.2.3.6 Retail and Consumer IoT

IoT devices such as smart shelves and beacons are FL-ready, as they can work locally to deliver customized customer experiences. There are other ways that retailers have to enhance product recommendations and store layouts of shops without compromising the information of their customers on a central database. In inventory management, FL is useful in IoT-enabled inventory systems to predict stock and scheduling of replenishment by training from data of other stores while keeping inventory data local optimally.

These applications show how FL enables IoT devices to utilize and perform complex ML computations while promoting data privacy, security, and sustainable optimization.

5.3 FUNDAMENTALS OF REINFORCEMENT LEARNING

RL is a type of ML in which an agent makes decisions based on the interactions with its environment. In contrast with supervised learning that involves a labeled dataset, RL gets trained from real-time outputs. The action that the agent takes is rewarded or penalized and based on this input the agent updates its decision policy. In RL the goal or the intended outcome is to increase the total of rewards over time, hence defining success or performance because of the intended goal of RL. The basic concepts of RL [18] are well understood by the following entities:

Agent: The RL decision-maker. By the function of rewards, the agent navigates in the environment and learns to do better.

External Environment: That is anything that the agent is in contact with. The activity of the agent alters the space: it either reinforces or penalizes, depending on the result.

State (s): An environment scenario or setting that has important implications for organizations and firms. This also means that the state has all the information concerning the decision-making of the agent.

Action (a): It means the actions or the possible choices an agent is capable of performing in a state. Acts exert impacts on the environment and the rewards of the adaptive module or agent. Following an action, the state provides the agent with a scalar reward denoted by (r). The reward indicates how beneficial or detrimental the action was in the context of the agent's goal.

***Policy* (π):** A policy is a plan the agent follows when carrying out his/her functions. A stochastic or deterministic strategy picks an action to take for each state through a probability distribution.

***Value Function* ($V(s)$):** It calculates the total benefit an agent anticipates he or she would gain when in a certain state with a particular policy. This allows the agent to examine the capacity of a state in the long run.

***Q-Function* ($Q(s, a)$):** The action-value function estimates the sum of future rewards starting from the current state and the action taken as well as the policy to be followed thereafter. Among all these action evaluation approaches in the field of RL, the Q-function comes out to be very crucial.

Exploration vs. Exploitation: Exploration is an approach where the agent attempts different actions in the environment to determine the reaction and be knowledgeable about the surrounding. Whereas exploitation is the phenomenon involved in selecting behavior knowing beforehand that this will lead to high returns in accord with the agent's experiences. Exploration and exploitation are actually a difficult dilemma in RL since the agent must learn through exploration but gain high utilities from exploitation.

Return: An agent's total compensation up to a particular time to episode termination. This means that there are tasks that give limited results while others produce endless results.

***Discount Factor* (γ):** The discount factor determines near-term reward against future benefits and can be 0 or 1. A discount factor near to 0 is highly tipped toward the instant reward, while "1" tipped toward long-term reward.

5.3.1 RL Challenges in IoT

The consequential problems related to the nonstationary behavior of the devices and sensors in IoT may be mitigated by using adaptive learning rate mechanisms, concept drift handling, and multi-agent RL techniques. The sensing capabilities of IoT devices are limited, which makes partial observability almost impossible, but partially observable Markov decision processes and belief-based approaches deal with such information. Coordinate learning protocols and decentralized and FL approaches can address the communication and coordination problems in distributed systems. While IoT devices contain modest or even negligible processing power and energy, such as lightweight RL algorithms, efficient and structured work offloading to centralized computing can be done. Security and privacy require safe computation, safe communication, and differential privacy.

Real-time decision-making in IoT needs RL algorithms to be computationally efficient and that can easily adapt transfer learning and simulation learning to enhance the learning process. Increased optimization and adaptive learning rates can help to increase the rates of recovery from rapid environment changes such as the failure in the network [19]. To achieve compatibility, there is a requirement for modular and interoperable solutions, including standard protocols as well as APIs, which can connect RL with older IoT frameworks.

5.4 CONVERGENCE OF 5G/6G AND IOT

IoT will be revolutionized globally by 5G/6G, which aims to combine the best features of both technologies. 5G and the future 6G will provide IoT devices the ability to share data at a high speed, low latency, and with increased connectivity. This combination is essential to enable future IoT applications, which include but are not limited to smart cities, self-powered vehicles, and Industry 4.0. For example, in smart cities, it is possible to enhance the effective management of city infrastructures with the help of 5G/6G IoT integration related to real-time traffic management [20], energy, and public safety. It can enable doctors to monitor patients remotely as well as administer telemedicine services with more reliability and speed, hence improving the accessibility and efficiency of medical solutions.

5.4.1 Overview of 5G and Upcoming 6G

5G networks are revolutionizing connectivity with their incredible speed and efficiency. With data transfer rates up to 10 Gbps, streaming ultra-high-definition videos and using immersive virtual reality and augmented reality applications is seamless. One of 5G's standout features is its ultra-low latency—down to just 1 millisecond—making it perfect for critical applications like autonomous cars, remote surgeries, and industrial automation. Its ability to connect up to a million devices in a single square kilometer makes 5G a game changer for dense areas like smart cities, where sensors and IoT devices abound. Network slicing is another breakthrough, allowing tailored virtual networks to meet specific needs, whether for gaming, healthcare, or emergency services [21]. On top of all this, 5G is designed to be energy-efficient, helping reduce the power consumption of connected devices, which is essential for building a sustainable, interconnected future.

Computation capability in 6G imagined for the period before the 2030s has been anticipated to offer much more advancement. A speed of 1 Tbps (terabits per second) would enable sub-millisecond latency for real-time AI applications, such as holographic communications and haptic internet. Self-optimizing networks are powered by AI and 6G that enable machines, environments, and human senses to interact through Brain-Computer Interfaces which reduce the emission of greenhouse gases, and enhance energy utilization efficiency to maintain sustainable development.

6G can be employed in applications such as tele-holography, smart city, remote surgery, diagnostics through AI, and self-driving cars. The issues that need to be considered are the availability of the spectrum, the development of the necessary infrastructure, security issues, concerns about privacy, and the need for applicability on a global level and the necessity of standardization on the international level. The transition from 5G to 6G will make the internet faster and will enhance the intelligence and sustainability of the entire world, which will definitely enable FRL in IoT.

5.4.2 Advancing IoT with 5G/6G

In general, 5G and 6G are more suitable for IoT applications due to faster speeds, low latency, and better device connectivity. Self-driving automobiles, industrial robots, and intelligent cities utilize 10 Gbps data throughput and one million devices per square kilometer. 6G will include ultra-reliable low-latency communication (URLLC) for new activities, including robot surgeries and AI applications and data rates of 100 Gbps. Increased device density, longer battery life, and secure networks will allow IoT applications such as brain-computer interfaces, holographic interface, smart grid, and smart transportation systems with the advent of 5G and 6G. The implementation of 6G will increase the number of connected devices without any human interference in the processes of smart manufacturing and the automation of supply chains.

5.5 INTEGRATION OF FL WITH RL IN IOT

Combining FL and RL within the IoT can lead to smarter, more decentralized, and privacy-conscious solutions. With FL, sensitive data such as user habits and environmental conditions stay on individual devices instead of being sent to a central server, safeguarding privacy [22]. Meanwhile, RL helps these IoT devices learn from their experiences, making decisions and adapting through trial and error based on past outcomes. When FL and RL work together, IoT networks can harness the collective knowledge of many devices while responding swiftly to real-time changes. This powerful combination helps to build complex, high-quality, and safe IoT applications in homes, cars, and factories to make immediate and accurate decisions locally.

5.5.1 Need for Integrating FL and RL in IoT Systems

One of the main issues arising from the enhanced digitization is the growing demand for FL and RL within the IoT system due to privacy, efficiency, and adaptability requirements. The large number of IoT devices yield huge amounts of sensitive data, which is processed by a central server; this causes privacy concerns. FL stores the data at the device level and does not reveal it to reduce the risk as it trains the machines at the devices. RL fundamentally enhances the versatility of the IoT system because the devices obtain the ability to enhance their performance based on real-time inputs. The deployment of IoT systems is a balance of data confidentiality and learning adaptability. FL and RL can be used to process intelligent and effective responses to complex scenarios. Those applications related to data-sensitive information and real-time decision-making, such as smart cities, healthcare, and autonomous systems, need this connection.

This combination also tackles various IoT system problems. FL prevents the data from being sent to a central server, which effectively takes care of privacy concerns [23]. The devices can make decisions based on the environmental feedback they receive through the method of dynamic and adaptive learning in RL. The FL and RL consolidation not only saves the bandwidth by generating and processing the data at the location but also the time of the process. Besides, FL divides the learning process into multiple devices, thus it makes life easier for the central servers, whereas RL also increases the decision of the single device. Finally, real-time problem-solving and resource optimization enhance the IoT systems with more stability and better application handling capabilities.

5.5.2 Benefits and Challenges of Combining FL and RL Techniques

FL and RL enable countermeasures against privacy, scalability, adaptive decision-making, and resource utilization in IoT systems. FL means data are stored on the devices themselves rather than rare central computers, which are more private and secure. Some examples of application of this strategy are in healthcare, finance, and smart homes where the privacy of data is crucial. The FL and RL methods defend personal data while assisting systems to learn from data if integrated. The integration of both of these elements is required for building the user's trust and being GDPR-compliant.

When it comes to more IoT devices, FL's approach to learning, which is decentralized, reduces the load on the central infrastructure and also can be easily scaled. When combined with a well-suited RL, this method enhances efficiency in scaling and the application of large-scale SC utilization, such as in smart cities and industrial automation, where the method is well suited for realistic application due to its continuous and adaptable learning characteristics. The FL enhances RL's real-time decision-making capacity to change IoT devices' conditions. FL saves resources like bandwidth and energy used in transferring data as it minimizes the number of data being transferred. Thus, efficiency is categorically required for sustainable funding in resource-scarce conditions. FL enhances the accumulation of knowledge related to the devices and breaks down RL's repetitive learning affording generalization across different environments.

In the same regard, the merging of FL and RL is also a complex process. Switching models often between devices can lead to elevated delay and lessened performance using rather frequent forces in large-scale or bandwidth-limited applications [24]. This data is therefore characterized by features such as heterogeneity and nonindependent and identical distribution and for this reason, the training of a model using IoT data presents itself as not only challenging but also inefficient. Because of the vulnerability of RL to noisy feedback and the challenges of combining models from various devices, the achievement of model synchronization and coherency is challenging. FL and RL are compute and power-hungry procedures that might surpass IoT devices' capabilities. Lastly, FL enhances privacy; however, it also increases extra security threats such as model poisoning attacks that hinder the speed of a system's performance. Both factors are still relevant to the improvement of FL and RL of the next-generation IoT applications.

5.6 ADVANCED TECHNIQUES AND ALGORITHMS

IoT technology has benefited from changes in 5G and 6G technologies as well as FRL. These enhanced communication and ML technologies have thus been especially important due to the increased number of IoT devices and the increased complexity of learning and decision-making required by each. From 5G to 6G, FRL systems can have faster and lower latency to enhance the capacity to regulate the complexity of today's IoT scenarios.

5.6.1 Advanced RL Techniques for Federated Learning

FL leads this revolution because it shifts the traditional approach of training ML models in a decentralized context such as IoT [25]. FL is enhancing the capabilities of IoT applications by employing the processing capability of distributed devices and maintaining data privacy. Table 5.1 outlines some of the most prominent techniques; contemporary FL techniques are always developing to enhance how devices cooperate and learn from one another.

5.6.2 Advanced RL Algorithms for Federated Learning

The combination of RL and FL is discussed in a scenario of distributed learning and different local contexts need to be taken into consideration. Superior algorithms to federated RL address these factors in numerous ways as illustrated below. The FedDQN algorithm adopts FL concepts into the typical Deep Q-Networks (DQN). RL agents can develop the maximum policies locally in addition to synchronizing and updating the global Q-values in this approach; it is useful for distributed device studying. The Federated Policy Gradient Methods help in the optimization of the decision-making policies by employing policy gradients. These solutions involve both regional possibilities and global approaches by relying on regional realities and often phasing with a global practice. The other technique is Asynchronous Actor-Critical Algorithms. As for Asynchronous Actor-Critic Agents (A3C), the federated setting regulates the asynchronicity of device updates. The above solution helps in dealing with federated contexts related to policy updates with regard to the concerns of delay and inconsistency [26]. Last but not least, Meta-Reinforcement acquiring improves the ability for fast application of the newly acquired knowledge in new tasks or different environments by acquiring meta-information from previous experience. This is particularly effective in FL since devices undergo several dynamic scenarios. These powerful RL algorithms helped to reduce the complexity of FL while enhancing distributed IoT learning systems' performance and resiliency.

5.6.3 Hybrid Approaches and Their Benefits

Incorporating two types of ML, FL and RL, into hybrid systems offers a number of advantages in improving IoT systems' learning potential and their performance. These hybrid methods make use of the capabilities of both FL and RL to provide the following benefits. First, they improve privacy and individuality features to a certain extent. FL stores information on local devices thus making sure that none of the sensitive information gets out, and RL adapts the system's reactions to the behaviors of the devices, thus creating unique yet personal experiences without compromising the user's privacy.

Furthermore, such a combination results in increased scalability as well as flexibility of the approaches. FL can also benefit IoT systems' large numbers of connected devices through the FL process of local model updating. In the same respect, RL capability enables devices to self-adapt to local conditions thus making RL a highly flexible system in that it can cope with different scenarios in an effective manner.

TABLE 5.1
State-of-the-Art FL Techniques in IoT

S. No.	Techniques	Key Features	Challenges	Use Cases in IoT
1	Federated Averaging (FedAvg)	Combines model updates from local devices by averaging weights.	Communication overhead Non-IID data (data may not be independently and identically distributed).	Smart homes Wearables in healthcare (e.g., activity recognition).
2	Federated Stochastic Gradient Descent (FedSGD)	Computes gradients on local data, updates model iteratively	Requires synchronization Slower convergence in edge environments.	Energy management in smart grids Intelligent transport systems.
3	Personalized Federated Learning (pFedMe)	Allows devices to maintain personalized models while contributing to global learning.	Balancing personalization and global learning.	Personalized healthcare (e.g., predicting health risks based on individual data).
4	Hierarchical Federated Learning (HFL)	Employs multiple tiers of aggregation (e.g., edge devices and cloud).	Increased complexity in the hierarchical model.	Large-scale IoT systems (e.g., smart cities), Multi-access edge computing (MEC) integration.
5	Asynchronous Federated Learning	Devices can contribute updates asynchronously, mitigating the issue of stragglers.	Heterogeneous hardware performance Security and privacy concerns with asynchronous updates.	Real-time monitoring in IoT networks Low-latency applications in healthcare or industrial IoT (IIoT).
6	Federated Transfer Learning (FTL)	Combines federated learning with transfer learning to leverage knowledge from similar tasks.	Domain shift issues Requires pretrained models.	Cross-domain IoT applications where data across different domains (e.g., healthcare and retail) can be utilized.
7	Robust Federated Learning	Designed to handle adversarial attacks, unreliable devices, and noisy data.	Higher complexity Requires robust aggregation algorithms.	Security-critical IoT applications (e.g., autonomous vehicles, industrial IoT).
8	Secure Aggregation Techniques	Focuses on privacy-preserving federated learning, such as homomorphic encryption and differential privacy.	Performance overhead due to encryption Complex key management.	IoT applications requiring stringent privacy, such as healthcare and financial systems.

Resource utilization is also beneficial, for FL, there is not a constant need for data transmission, while RL makes the best out of decision-making possibilities for devices respecting their resources.

The integration also enhances the resiliency of IoT systems and the resilience of IoT systems in general. Analyzing FL's aggregation techniques, one can note that problems concerning unstable or even malicious updates are solved to a certain extent, whereas RL improves the system's adaptability and stability in the face of questions and unpredictable changes [27]. Hence, the hybrid models make provision for adaptive learning and intelligent decision-making processes to enhance the ability of devices to improve their performance based on events and information collected within a local context and those gathered from a global context. The fast learning process makes it possible for the IoT systems to adapt to different conditions and user requirements thus coming up with improved and even more responsive applications.

5.7 CHALLENGES AND FUTURE OPPORTUNITIES

5G/6G-enabled FRL in IoT contexts brings several opportunities and challenges. Understanding these issues is essential for guiding future research and development.

5.7.1 Challenges

The integration of FRL in IoT infrastructure has some major issues. Two major problems that are evident because of heterogeneity and scalability: first, the sheer variability of smart devices with different computational power and limited resources. This aspect has made the scaling of FRL difficult and also ensured that there is a balanced performance and contribution to learning. Other challenges raised by the incurred communication overhead include the fact that the constant updating of IoT devices as well as the central servers tends to be an overwhelming workload on the network, hence causing delays in larger systems. There are also threats that relate to data privacy and security since transmitting model updates can be intercepted or manipulated, threatening sensitive applications and compromising trust.

Another limitation, which may affect many IoT devices, is the resource limitation in a device, including battery, computational intensity, and storage space, which may hamper the execution of complex RL algorithms [28]. The timing is of essence for applications utilizing real-time information as the time taken to aggregate the models and update them will slow down the system's performance. Algorithmic complexity comes from the fact that it is challenging to design FRL algorithms with good exploration/exploitation capability and the capability of running properly in decentralized and dynamic environments. Finally, the ethical and regulatory issues are important as the implementation of the AI-based systems working with the sensitive data should take into consideration the data protection laws and possible issues with the fairness and bias of the system, which may lead to legal problems or loss of users' trust.

5.7.2 Future Opportunities

To push forward FRL for IoT systems, algorithms need to be of flexible learning type which is capable of learning from the existing heterogeneous devices. Hierarchical FL and federated meta-learning should be the two major areas of research to develop efficient and effective FRL solutions for large-scale networks. It is also essential to enhance communication efficiency by utilizing such methods as quantization, sparsification, and edge computing to minimize the data transmission load. The improvement of privacy and security, through robust approaches like differential privacy along with the integration of blockchain-based reference models, will play a crucial role in safeguarding FRL processes and detecting malicious strategies.

Due to the limitations of IoT devices, algorithms are required that consume less energy; for the future works, RL models with low energy complexities and energy-color-optimal scheduling can be considered [29]. Low-latency solutions are always needed for real-time applications and help could be sought from developments in 6G technology. Industry-standardization for FRL implementation and concern for ethical AI with regard to enterprise compliance will optimize FRL application. Extending FRL toward many different areas and linking it with quantum computation could bring new ideas and enhance machine's scalability. Solving these issues will enhance the capability of 5G/6G for FRL in IoT to increase the dependability and efficiency of the systems and make it possible to implement in various scenarios.

5.8 CONCLUSION

FRL under the 5G/6G IoT has enhanced smart systems' performances and use cases. FRL has upgraded with the 5G/6G implication to enhance privacy, real-time decision, and scalability within various and constrained IoT environments. The first one, FL, prevents the exposure and identification of a user's personal data to third parties since data is not uploaded to a central server for training while the second one, RL, has the capability of learning through experience and making intelligent decisions as soon as feedback is provided. Altogether these technologies, with 5G/6G networks' ultra-low latency and high bandwidths, guarantee real-time, auto/adaptive IoT systems.

The integration of FL–RL has a lot of significance for the IoT. It empowers systems more intelligent and more adaptive and will be able to scale large amounts of data in a small amount of time. The inclusion of such measures would pave the way to driverless cars, better industrial robotics, and smart health systems among other fields while conforming to editorial privacy standards of different countries, changing several sectors. These applications can be adapted and optimized in real time to enhance their efficiency, user experience, and operation results. In addition to the above, business and academia can leverage 5G/6G to enhance the development as well as the implementation of the FRL system. They will positively impact the industry's performance, generate new business prospects, and enhance IoT products. Organizations provide academics with a broad ground for algorithm research, privacy research, and exciting application research. Cognitive, safe, and effective IoT systems are achievable if researchers and practitioners provide solutions to the problems affecting the design and implementation of IoT systems.

REFERENCES

1. Kairouz, P., McMahan, H.B., Avent, B., Bellet, A., Bennis, M., Bhagoji, A.N., Bonawitz, K., and Charles, Z., Advances and open problems in federated learning, *Foundations and Trends in Machine Learning*, 14(1), 1–210, 2021.
2. Ande, R., Adebisi, B., Hammoudeh, M., and Saleem, J., Internet of Things: Evolution and technologies from a security perspective, *Sustainable Cities and Society*, 54, 101728, 2020.
3. Yogeshwari, M., Prasanth, A., Open research challenges in wireless technologies case study, *Wireless Communication Technologies – CRC Press*, 301–317, 2024.
4. Guo, F., Yu, F.R., Zhang, H., Li, X., Ji, H., and Leung, V.C.M., Enabling massive IoT toward 6G: A comprehensive survey, *IEEE Internet of Things Journal*, 8(15), 11891–11915, 2021.
5. Dung, N.T., Thang, P.C., and Trang, T.T.N., Convergence of IoMT and blockchain for emerging personalized healthcare system: Challenges and use cases, *Technological Advancement in Internet of Medical Things and Blockchain for Personalized Healthcare*, 99–109, 2024.
6. Kober, J., Bagnell, J.A., and Peters, J., Reinforcement learning in robotics: A survey, *International Journal of Robotics Research*, 32(11), 1238–1274, 2013.
7. Mnih, V., Kavukcuoglu, K., and Silver, D., Human-level control through deep reinforcement learning, *Nature*, 518(7540), 529–533, 2015.
8. Aruchamy, P., Lavina B., and Devi, S. An energy-aware link fault detection and recovery scheme for QoS enhancement in Internet of Things-enabled wireless sensor network, *Computers and Electrical Engineering*, 123, 1–17, 2025.
9. Yang, Q., Liu, Y., Chen, T., and Tong, Y., Federated learning: Challenges, methods, and future directions, *ACM Computing Surveys (CSUR)*, 1–21, 2019. https://doi.org/10.48550/arXiv.1908.07873.
10. Yu, C., Liu, S., Lai, L., and Liu, Y., Federated reinforcement learning: Fundamentals, algorithms, and applications, *arXiv preprint*, arXiv:2202.09025, 1–39, 2022. https://doi.org/10.48550/arXiv.2108.11887.
11. Simsek, M., Aijaz, A., Dohler, M., Sachs, J., and Fettweis, G., 6G wireless networks: Vision, requirements, challenges, insights, and opportunities, *IEEE Communications Magazine*, 109(7), 1166–1199, 2020.
12. Giordani, M., Polese, M., Roy, A., Castor, D., and Zorzi, M., Towards 6G networks: Use cases and technologies, *IEEE Communications Magazine*, 58(3), 55–61, 2020.
13. Aledhari, M., Razzak, R., Parizi, R.M., and Saeed, F., Federated learning: A survey on enabling technologies, protocols, and applications. *IEEE Access*, 8, 140699–140725, 2020.
14. Niknam, M., Dhillon, H.S., and Reed, J.H., Federated learning for Internet of Things: Recent advances, taxonomy, and open challenges, *IEEE Communications Surveys & Tutorials*, 23(3), 1759–1799, 2020.
15. Yang, K., Jiang, T., Shi, Y., and Ding, Z., Resource-constrained federated learning: A comprehensive survey, *IEEE Transactions on Wireless Communications*, 23(3), 1622–1658, 2021.
16. Pamarthi, S., and Narmadha, R., Literature review on network security in wireless mobile ad-hoc network for IoT applications: Network attacks and detection mechanisms, *International Journal of Intelligent Unmanned Systems*, 10(4), 482–506, 2022.
17. Devakirubai, N., Velu, A., and Sumathi, D., Optimizing vehicle-to-vehicle (V2V) communication efficiency with KNN-based dynamic time slot allocation, *Proceedings in*

2024 International Conference on Social and Sustainable Innovations in Technology and Engineering, Sasi Institute of Technology and Engineering, Tadepalligudem, Andhra Pradesh, India, 1–6, 2024.

18. Li, H., Ota, K., and Dong, M., Learning IoT in edge: Deep learning for the Internet of Things with edge computing, *IEEE Network*, 32(1), 96–101, 2018.
19. Chen, C., Hong, X., and Wei, W., Communication-efficient federated learning with adaptive parameter freezing, *IEEE 41st International Conference on Distributed Computing Systems (ICDCS), Washington, DC, USA*, 1–11, 2021.
20. Morgado, A., Huq, K.M.S., Mumtaz, S., and Rodriguez J., A survey of 5G technologies: Regulatory, standardization and industrial perspectives, *Digital Communication Network*, 4(2), 87–97, 2018.
21. Balraj, L., Prasanth, A., Sowndarya, K., and T. Kuntavai, A lightweight blockchain scheme for secure data communication in the Internet of Things-enabled wireless sensor network, *Proceedings of IEEE International Conference on Smart Systems for Applications in Electrical Sciences*, Siddaganga Institute of Technology, Tumakuru, Karnataka, India, 1–6, 2024.
22. Pinto, E.C., Sadeghi, S., Zhang, X., and Dadkhah, S. Federated reinforcement learning in IoT: Applications, Opportunities and Open Challenges, *Applied Sciences*, 13(11), 6497, 2023.
23. Liu, W., Xu, X., Wu, J., and Jiang, J., Federated meta reinforcement learning for personalized tasks, *Tsinghua Science and Technology*, 29(3), 911–926, 2024
24. Qi, J., and Qihao, Z., Federated reinforcement learning: Techniques, applications, and open challenges, *Intelligence & Robotics*, 1, 1–15, 2021.
25. Qiang, Y., Yang, L., and Tianjian, C., Federated machine learning: Concept and applications. *ACM Transactions on Intelligent Systems and Technology*, 10(2), 12, 2019.
26. Shafi, M., Andreas F., and Peter J., 5G: A tutorial overview of standards, trials, challenges, deployment, and practice, *IEEE Journal on Selected Areas in Communications*, 35(6), 1201–1221, 2017.
27. Yange, C., Lei, L., and Yuan, P., A privacy-preserving federated learning framework with lightweight and fair in IoT, *IEEE Transactions on Network and Service Management*, 21, 5843–5858, 2024.
28. Yao, J., and Sun, X., Energy-efficient federated learning in internet of drones networks, *2023 IEEE 24th International Conference on High Performance Switching and Routing (HPSR), Albuquerque, NM*, 185–190, 2023.
29. Zhaohui, Y., Mingzhe, C., Kai, W., Federated learning for 6G: Applications, challenges, and opportunities, *Engineering*, 8, 33–41, 2022.

6 Blockchain-Integrated Federated Learning for IoT-Based Smart Applications

S. Balakrishnan, Syed Shahul Hameed, Sunil Kumar RM, and S. Simonthomas

6.1 INTRODUCTION

6.1.1 Background

Over the past few years, the Internet of Things, or IoT for short, has transformed the generation and usage of data in all industries, from healthcare to smart cities, and industrial automation to the farthest extent. Because the nature of IoT-based systems is profoundly decentralized, they introduce complex challenges, especially in terms of data privacy, security, and efficiency of processing [1]. To address the above issues, an emerging solution that combines the advantages of federated learning (FL) and blockchain technology is brought up in this chapter, namely blockchain-integrated federated learning (BIFL). FL is a new paradigm, which allows multiple machine learning models to be learned collaboratively across distributed devices, in the form of sensors and smartphones, with no raw data transfer to central servers [2]. This approach preserves data privacy as it aggregates updates that are computed locally rather than over networks sharing the sensitive information. Though FL effectively solves issues due to privacy concerns associated with decentralized data sources, blockchain will find an application for this framework by having a secure, decentralized ledger [3].

The features of blockchain, namely, immutability, transparency, and cryptography security properties it develops, are considered an ideal framework to "improve the trustworthy and integrity level of federated learning in IoT environments". While utilizing blockchain, BIFL ensures that all transactions and interactions within the FL process are recorded tamper-resistantly, hence bringing in utmost changes to risks connected with data manipulation and unauthorized access [4]. BIFL presents new incentive mechanisms developed around smart contracts that provide monetary incentives to IoT device owners to take part in FL tasks by rewarding contributions toward model training with no invasion of sovereignty over data. These incentives

DOI: 10.1201/9781003532323-6

enhance collaboration on data among participants, thereby creating a more robust and inclusive IoT-based smart applications ecosystem [5]. Blockchain with FL integrated into IoT environments demonstrates a collaborating advantage that greatly enhances data privacy, security, and scalability. Case studies and theoretical analyses done by us show that BIFL can transform industries through its secure, efficient data-driven decision-making processes that respect the rights of privacy of individuals and organizations involved within the IoT networks [6].

6.1.2 Motivation

The inspiration to design the BIFL comes from the challenge posed by IoT-based smart applications regarding data privacy, security, and efficient processing of the data [7].

Classical aggregation and analysis of data in the IoT scenarios usually have problems with the following:

1. Data Privacy Concerns: IoT devices tend to produce sensitive data, and if this kind of information were to become exposed, it would be harmful not only to the privacy rights of individuals but also organizations' confidentiality. Centralized data aggregation models will face a considerable risk of breaching data and unauthorized access.
2. Data Security Risks: IoT data storage and processing in centralized hubs are considered a cyberattack prone hub that raises questions about data integrity and system reliability. The increasing number of mounted IoT devices makes it extremely challenging to secure large amounts of IoT data.
3. Scalability Problems: Centralized infrastructures just cannot scale to keep pace with increasing data and computing capacities as the IoT networks grow. Scalability can affect the deployment of real-time, data-driven applications across many IoT ecosystems.
4. Unreliable Trust and Lack of Transparency: IoT ecosystem stakeholders such as device owners, data analysts, and end-users lack transparency in the handling or model training process for the data. This risk of nontransparency threatens to challenge effective cooperation and innovation opportunities in IoT applications.

BIFL addresses these challenges by combining the strengths of FL and blockchain technology [8].

Enhanced Data Privacy: FL allows IoT devices to learn together on machine learning models without ever sharing raw data, thus providing strong privacy by preservative at source. In the blockchain, the data privacy rights are protected through open recording and secure management of all the interactions and transactions that occur in the FL process.

Enhanced Security: The decentralized architecture along with cryptography for the blockchain will provide a safe mechanism to store aggregated model updates while ensuring integrity in data across IoT networks. This will greatly boost the defenses against cyber threats and against unauthorized data manipulation.

Scalability and Efficiency: BIFL mitigates all the scalability problems that are natural for a centralized infrastructure because of task and storage delegation over a decentralized network of IoT devices. This also enables real-time edge processing and analysis that make responses much faster while optimizing the use of resources.

Transparency and Trust: Transmutation into blockchain owing to transparency and non-amendable nature builds up great trust toward stakeholders as the data transaction record and activities at the training model are verifiable. Incentive mechanisms and agreements can be automatically enforced through smart contracts placed on blockchain, thus effective and successful collaboration and participation in FL tasks.

BIFL is one among the transformative applications of blockchain technology and FL for the architecture of IoT-based smart applications [9]. Since BIFL overcomes the challenges in terms of privacy and security, scalability, and trust, therefore, more resilient, transparent, and innovative IoT ecosystems are enabled, which helps to implement data-driven solutions effectively throughout various sectors with confidence [10].

BIFL for IoT-based smart applications enhance privacy, security, scalability, and promote transparency and trust by implementing incentives for diversified applications and allowing research and development [11]. BIFL is intended to help unlock the entire potential of IoT data that addresses challenges critical enough to empower stakeholders in the digital economy.

6.2 OVERVIEW OF FEDERATED LEARNING

6.2.1 Definition and Overview

FL is one of the new decentralized paradigms aiming toward collaborative model training on millions of "edge devices, such as smartphones and IoT devices and computers, without transferring the raw data to a central server". This eliminates one of the major problems concerning data privacy associated with centralized data storage and processing since it stores the data locally where it was generated [12].

Overview

1. Distributed Training: Distributed training is the process in which traditional machine learning is compared where data needs to be gathered at one central location for training models. Decentralized training goes around on individual devices or nodes belonging to a network. Using their local data, all devices work together to compute updates to a shared global model that usually happens by repeated rounds of model training and aggregation of updates.
2. Privacy-Preserving: FL applies the rationale behind it – keeping data decentralized and local. FL returns only model updates, often gradients, to the

central model instead of the raw data. It reduces the related risks that mean exposure of the information or sensitive data involved within the transfer procedure.

3. Collaborative Learning: FL devices collaboratively learn and improve a common model but without explicitly sharing local data. The model becomes especially valuable when data privacy regulations, for example, such as General Data Protection Regulation (GDPR), prevent data movement or sharing across jurisdictions.
4. Iterative Process: FL carries out an iterative process of rounds of communication between the central server and devices. For each round, devices have to compute model updates locally on their data and aggregate it at the central server to update the global model for subsequent application. It runs repeatedly till the convergence criterion is satisfied, or a certain number of cycles have been performed.

Overall, FL transforms the traditional paradigms of machine learning into self-improving models, which one can then start training collaboratively without sacrificing the privacy and security of data [13]. It permits edge devices to contribute their global intelligence without having to sacrifice secrecy over sensitive information and provides new avenues in decentralized and privacy-preserving artificial intelligence (AI) applications [14].

6.2.2 Architecture

The architecture of FL suggests several key components and interactions between these components, by which decentralized model training is allowed to be performed across distributed edge devices. The architecture of FL is shown in Figure 6.1.

Here's a more detailed breakdown of the FL architecture:

1. Centralized Coordinator (Central Server):
   Aggregator: It is another node which can coordinate, host, and manage the FL process. It can start and manage the communication with the participating edge devices.
   Global Model: The central server must maintain the global model, which essentially forms the very basic structure of the FL mechanism. FedAI updated the global model iteratively by contributions from the edge device.
2. Edge Devices (Clients):
   Data Owners: The owners of edge devices can be mobile phones, IoT devices, or computers with ownership over a local dataset and employing such datasets to train their models.
   Local Model Training: Each edge device performs local model training based on data at that device, and from these, updates of the model are generated

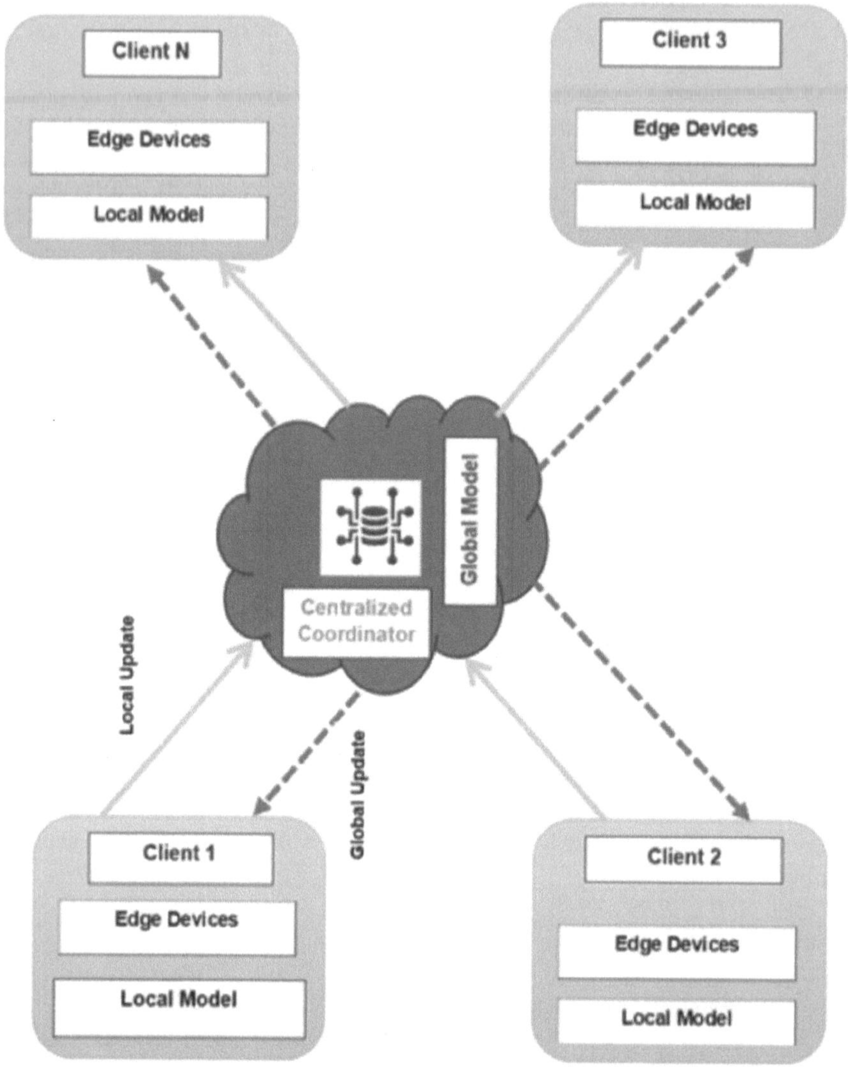

FIGURE 6.1 Architecture of federated learning.

through local procedures, mainly gradients or weights adjustments, which reflect local patterns and characteristics.

3. Communication Protocol:

Round-Based Communication: FL is a round or epoch-based system. During every round, the central server broadcasts the dominating global model to the chosen set of edge devices.

Update Model Transmission: The edge devices perform model updates locally on their data and transmit them to a central server.

Aggregation: This middle node aggregates the updates that are derived from models and updates global models. The aggregation might be average, weighted average, and so on in order to collect the different types of updates arising from different devices.

Thus, this architecture of FL combines the idea of central coordination together with decentralized execution on edge devices. Therefore, it provides cooperative model training while keeping data privacy and security safe. This makes it the suitable solution to a wide range of applications in healthcare, IoT, finance, and so on.

6.3 IOT IN SMART APPLICATIONS

6.3.1 IoT Overview

The IoT is the transformative power in the world of smart applications. It changes the general nature of conducting various sectors and services [15]. Essentially, the IoT connects ordinary objects and devices to the internet and provides for the collection, exchange, and application of data in previously unimaginable manners. From smart homes and cities to industries, health, agriculture, and anything beyond this, it ranges from interconnecting networks of devices and allows for smooth communication and automation [16].

In smart applications, IoT plays an important role to enable devices to collect real-time data, monitor environments, and respond sensibly to their changes. It has been a way that increases efficiency and productivity in enhancing new opportunities for innovation and improvement of user experiences. However, smart applications driven by IoT would revolutionize healthcare for remote patient monitoring, though they would optimize energy consumption in buildings [17]. This introduction provides a background for the exploration on how IoT empowers smart applications in every sector with promises pointing toward a future in which all aspects of living, working, and interacting with the surroundings can change.

6.3.2 Applications in Smart Environments

The IoT [18] changed the paradigm of smart applications in all walks of life by the capability of devices to collect and share and even respond to that data. In smart applications, IoT provides the following:

1. Smart Home Automation: Thermostats, lights, and security systems are joined to central hubs, which can be programmed and automated from anywhere with the help of choices or conditions in the environment.
2. Healthcare: IoT sensors track the patients' location from the remote areas, monitor the vital signs, and then update the data to healthcare professionals for better preventive care and timely intervention.
3. Industrial IoT: Manufacturing industries work upon the data analytics with the help of IoT technologies to know about the predictor maintenance and check on operations in real time.

4. Smart Cities: IoT sensors monitor traffic flow and consumption of energy along with waste disposal to work efficiently in the cities.
5. Agriculture: IoT will help promote precision farming while monitoring the status of soil moisture, weather condition, and crop health in order to optimize irrigation and pesticides utilization.
6. Retail: It supports doing personalized shopping, inventory management, and supply chain optimization through real-time analytics over the data.
7. Transportation: IoT transformed this field of logistics, fleet management, and passenger safety in regard to on-the-go analytics over GPS tracking, predictive vehicle maintenance, and traffic conditions.

In general, the use of IoT for smart applications does improve efficiency and reduces costs due to integral devices and insights derived by data.

6.3.3 Challenges and Opportunities

6.3.3.1 Challenges of IoT in Smart Applications

Security Concerns: Many devices in IoT have limited power, making it challenging to provide strong security measures. Therefore, they become highly vulnerable to cyberattacks and data breach.

Interoperability: Devices from different vendors are often other-based or incompatible protocols, which would make communication and integration more difficult.

Scalability: It is difficult as well as resource-intensive to operate very large IoT deployments and networks.

Data Privacy: This whole process of gathering and processing a huge amount of data also raises big issues in terms of user privacy as well as regulatory compliance.

Reliability and Maintenance: The IoT gadgets should be provided with reliable connectivity together with continuous maintenance so that the movement works efficiently.

Power Consumption: Many IoT devices are battery operated; hence, they require efficient power management strategies in order to increase the lifetime of the battery.

6.3.3.2 Opportunities of IoT in Smart Applications

Efficiency and Automation: IoT offers a simple way to enable automatic processes, reducing or optimizing resource usage and minimizing operational cost.

Smart Decision-Making: IoT generates timely data from IoT devices in the manner in which it claims efficiency for organizations that take decisions with an informed perspective to work on efficiency. It specifically has very advantageous benefits in retail, healthcare, and home technologies because it highly enriches the experience of the user.

Innovation and New business Models: IoT is encouraging innovation because it creates birth for new products, new services, and newer business models.

Sustainable Practices: IoT is as such propitious for sustainability in terms of energy management, an intelligent transportation system, and precision agriculture.

The Remote Monitoring and Control: The online equipment monitoring, predictive maintenance, and safety protocols make the industries both efficient and safe.

In response to those challenges, optimum use of opportunities may unlock the fuller potential of IoT in application innovation and contribute to social impacts in various industries.

6.4 BLOCKCHAIN TECHNOLOGY

6.4.1 Introduction to Blockchain

Among these, blockchain technology has emerged as the next innovation that could change the face of an industry, be it finance and healthcare or supply chain management and, more broadly, many sectors [19]. In simple terms, blockchain technology is "a decentralized digital ledger in which transactions are recorded across a network of computers in a secure, highly transparent manner". This brings a twist from the traditional centralized systems that borrow all functions from a controlling authority that also oversees data and transactions. Instead, blockchain-based applications are built on a distributed ledger wherein every participant within "the network has a copy of the ledger". Blockchain technology has changed the corporate landscape of trust, security, and transparency vis-à-vis digital transactions entirely [20].

The blockchain first saw development in supporting Bitcoin but has since evolved into so much more, becoming far bigger and broader. Its most unique features include irreversibility, cryptographic security, and mechanisms to reach consensus in two-way exchanges between parties so as to allow for direct transactions without mediaries. As blockchain becomes ubiquitous and permeates almost every other section within the economy, they promote cost reduction, streamlined processes, and elimination of risks. The introduction goes on to describe how blockchain technology works, its core components, and all the ways it is changing the shape of industries and reshaping digital ecosystems around the world [21].

6.4.2 Features and Characteristics

Blockchain technology was designed with some key features and characteristics that separate it from preceding centralized systems [22]. Main features of the blockchain comprise the following:

1. Decentralization: Blockchain actually comes to the form of a decentralized network of nodes on a computer. The underlying data forms and transaction operating are done through consensus among participants. It eliminates, as a result, having a central governing authority – the chances for manipulation or even deceit are reduced with more transparency.
2. Immutability Ledger: Any and all information placed in the blockchain cannot be deleted or changed in any way. Inside a blockchain, each block includes a

hash from the previous block to ultimately form a definite and time-ordered ledger of all activities.

3. Transparency: The blockchain ledger is transparent to parties in the network. Even when identities can be pseudonymous, transaction details usually are accessible for the build of trust and transparency.
4. Security: Blockchain makes use of cryptographic algorithms for ascertaining secure transactions and data establishment. There are mechanisms such as PoW or PoS for verification of transactions; hence, the job is almost impossible for some malicious actor to compromise the system.
5. Smart Contract: Ethereum blockchain introduces smart contracts as a software code self-executing, having predetermined rules and conditions with the ability to automatically go on to execute when conditions are met facilitating automated and trustless agreements.
6. Distributed Consensus: The algorithms for consensus give guarantee that all participants in the network have a collective agreement on the legitimacy of transactions. Three key categories of consensus mechanisms are available, characterized by their security level, scalability, and energy utilization level.
7. Tokenization: Using blockchain, one can create and transfer digital assets, including a token that signifies ownership or rights for the access to certain resources. They may be currencies, assets, or utility for a decentralized system.
8. Audibility and Traceability: With the help of transparency and immutability, blockchain has been able to facilitate relatively easy auditing of transactions and traceability of assets through all stages of life. The benefits are not exclusive to managing the supply chains, auditing, and being ensured of one's conformance to regulatory standards.

All these enablements have transformed the blockchain technology into a very valuable tool, which is to be developed with trust building and new business models and by revolutionizing industry fronts by providing secure, transparent, and efficient digital transactions and data management [23].

6.4.3 Use Cases of IoT with Blockchain Technology

With blockchain technology and IoT, several interesting use cases exist that enhance security and transparency as well as provide efficiency in many applications [24]. The most important and prominent highlights about use cases of blockchain in IoT include the following:

1. Device Identification and Authentication: This can be applied to have secure registration and management of identities of IoT devices on blockchain that can enable a unique cryptographic identity to each, thus enhancing authentication, hence preventing any unauthorized access or tampering.
2. Data Integrity and Security: In the way that this blockchain technology provides data integrity and security of IoT-data in the form of an immutable

record of transactions of data that ensure authenticity and provenance of data, thus being of paramount importance for healthcare patients' data, supply chain product traces, and smart city sensor data.

3. Decentralized Data Sharing: In a blockchain-based decentralized network, IoT devices share data securely with permissioned parties. With IoT devices sharing data, there is promise of data privacy; besides, dependence on such centralized servers that are prone to breach chances is diminished.
4. Automated Smart Contracts: Prefixed rules in blockchain-based smart contracts can be developed so as to interact and carry out transactions of IoT devices in an automated manner. This can, for example, be such that a logistics support system within a supply chain, where the smart contract automatically pays or delivers goods once the conditions of delivery monitored by IoT sensors are accommodated.
5. Supply Chain Transparency: Blockchain increases supply chain transparency and traceability as every transaction, such as the movement of products from origin to an end consumer, is logged. IoT sensors can monitor the time and location of the products where temperature and conditions are involved, all of which are recorded onto the blockchain in real time and allow for auditable transparency.
6. Energy Management: With blockchain-enabled IoT devices, energy consumptions in smart grids and buildings can be optimized. Devices can autonomously negotiate prices for energy, trade excess energy P2P, and transparently track the blockchain for the energy usage.
7. Asset Tracking and Management: Blockchain ensures correct, transparent tracking of physical assets using IoT sensors for the whole life cycle process. The application here would be priceless in logistics, for the healthcare industry tracing of medical devices, and in the machinery manufacturing industries.
8. Insurance and Risk Management: IoT data, integrated with blockchain, can impact the entire system of insurance. Immediate or real-time asset conditions related to vehicles or property can be provided for risk examination. Smart contracts, facilitating claims based on the satisfaction of smart conditions, reduce fraud and overhead cost.
9. Decentralized Autonomous Organizations: The blockchain and IoT just find an excuse to form decentralized organizations where IoT- based devices automatically decide and act upon predefined rules, governance protocols, etc.
10. Smart Cities: Blockchain-enabled IoT solutions can help smart cities through improvement in public services, optimization of waste management and flow of traffic, and by ensuring that the data collected about citizens of these cities through these sensors is private and secure.

Examples as such will also demonstrate the synergies that are involved when blockchain technology and IoT come together toward secure, efficient, and transparent systems across many different domains and can open innovation applications as well as business models [25].

6.5 INTEGRATION OF BLOCKCHAIN AND FEDERATED LEARNING

6.5.1 Motivation for Integration

Blockchain integration with FL will combine the strengths of two transformational technologies and address distinct kinds of challenges related to data management and privacy in the digital age [26, 27]. For these reasons, here are the motivations for integrating blockchain with FL.

1. Data Privacy: FL encourages training models on decentralized data but not by sending raw data – that is, it preserves the privacy of user data. The privacy of FL is further enhanced by using blockchain technology: safeguarding the transaction to manage permission for data access and ensure both integrity and transparency in the data.
2. Immutable Audit Trail: The application of blockchain technology results in a permanent ledger of data transactions. Blockchain technology introduces more transparency and accountability into FL systems, which means that data owners, model developers, and validators trust the stakeholders because authenticity and integrity are verified for data used for training purposes.
3. Decentralized Governance: Blockchain facilitates decentralized governance models in which individuals together control access to the data and updates of the model. Smart contracts implement the automation of FL network decentralized governance rules, ensuring an unbiased and transparent environment for its network.
4. Data Monetization and Ownership: The use of blockchain technology will enable making of transactions transparently and traced, thereby particularly on data ownership and usage rights. Data contributors under FL can safely monetize their data contributions in a controlled manner while holding access and usage control through the aid of blockchain-based smart contracts.
5. Secure Model Deployment: The blocks would provide secure deployment and execution of FL models across the distributed nodes. Smart contracts will help in maintaining the updates of the model components with validation of consent among the participating nodes and thus enforce the data privacy regulation at all stages of model life cycle.
6. Anti-attack: With its decentralized architecture and using cryptographic security, blockchain is much more robust against FL networks being the victims of malicious attacks, data breaches, and tampering attempts. Sensitive data and models within the FL network can only be protected by an appropriate security framework.
7. Scalability and Interoperability: Blockchain can particularly utilize scalability solutions that involve layer two protocols and sharding for enabling good performance in FL networks that have large data distributed among numerous IoT devices, edge computing nodes, and cloud servers.
8. Compliance to Regulation: Block chain transparency and auditability helps prevent noncompliance with data protection regulations such as GDPR as

well as the industry standards for FL deployments. Assurance of compliance is crucial for commercialized enterprises and organizations that are operating in very regulated industrial sectors.

Generally, blockchain and FL at the same time overcome the major pain points and open completely new opportunities for data privacy, transparency, governance, and security in distributed networks for FL models of AI being developed collaboratively and under conditions of privacy across decentralized networks [28].

6.5.2 Architectural Framework

Designing strong architecture frameworks that use the strengths of both technologies to make the realization of secure, privacy-preserving, and efficient AI model development in distributed networks will be the essence of the integration of blockchain with FL. There is an integrated architecture framework, as displayed in Figure 2, that encompasses the integration of blockchain with FL:

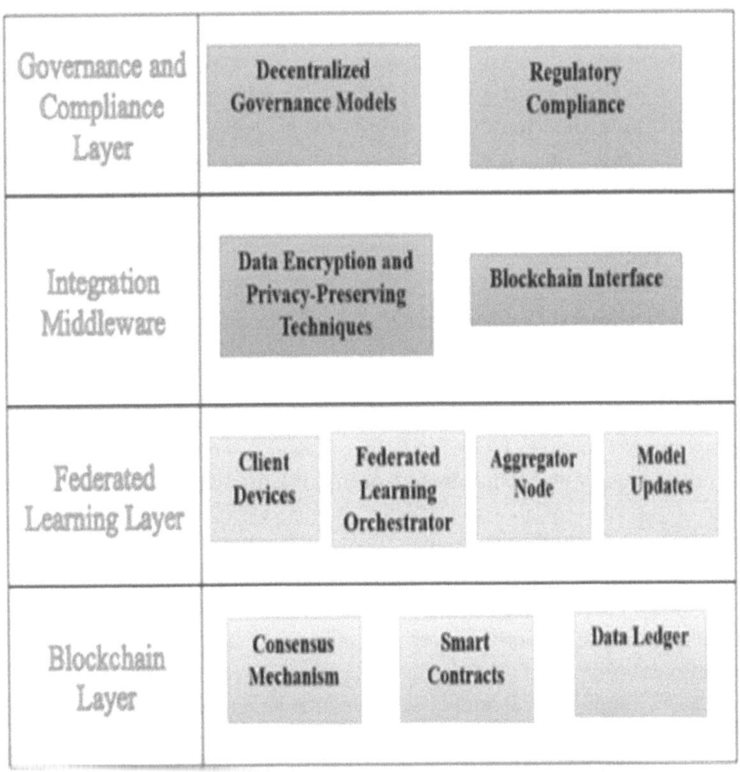

FIGURE 6.2 Architecture of blockchain and federated learning.

6.5.2.1 Components of the Architecture

Blockchain Layer

Mechanism for Consensus: This explains how nodes agree over the state of a blockchain. Typically, it is either proof of work or proof of stake, and in the case of FL, it may take a variant such as Federated Byzantine Agreement, or FBA.

Smart Contracts: Smart contracts determine the operating rules that describe the governance structure to access data, use models, and dispense incentives in the FL network.

Data Ledger: This stores hash of metadata of datasets and training parameters for providing data integrity without exposing raw data along with traceability.

Federated Learning Layer

Client Devices (Nodes): IoT devices and edge nodes among many distributed systems that offer local data for the purpose of model training without disclosing the underlying raw data.

Federated Learning Orchestrator: It organizes model aggregation, validation, and deployment across involved nodes while paying respect to privacy constraints.

Aggregator Node: It aggregates and collects updates coming from the models of participating nodes; no raw data is left to leave local devices.

Updating and Validation of Models: It sends model updates encrypted securely and validates those updates via consent mechanisms ensuring model integrity.

Integration Middleware

Data Privacy Preservation Mechanisms: Applying differential privacy and homomorphic encryption on the side of model training and aggregation as protection against threats to data privacy.

Blockchain Interface: It provides an integration interface between the FL components, and the blockchain layer assures appropriate interaction with security integrity and transparency, promoting trustworthiness and security.

Governance and Compliance Layer

Decentralized Governance Models: Smart contracts specify access rules for data, participant roles, incentives, and mechanisms of dispute resolution throughout the FL network.

Regulatory Compliance: It simply adds that the development of the AI model is in line with data protection regulation and also in line with industrial standards and ethical practices.

6.5.2.2 Workflow

Initialization and Registration

The nodes start by registering themselves on the FL network through publishing metadata with cryptographic proofs on the blockchain.

The eligibility of nodes is evaluated, based on specific eligibility criteria by smart contracts depending on factors like data quality and computational capabilities.

Model Training and Aggregation

The orchestrator assigns a global model to the nodes participating.

These nodes undertake the local training of the model using their private datasets, and therefore, these nodes will generate encrypted updates of the model.

The aggregator node gathers encrypted updates, aggregates them securely, and then moves the updated global model to the relevant nodes.

Consensus and Validation

Mechanisms of Blockchain Consensus: For the verification of integrity and authenticity of updates toward the model, blockchain consensus mechanisms are responsible.

Smart contracts make validation testing to ensure that individuals or groups adhere to the predetermined regulations of their minimum participation thresholds as well as quality benchmarks.

Incentive Mechanisms

Smart contracts will allocate incentives; tokens and rewards will be given to nodes according to their contribution in model training, as well as the quality of data provided, together with implications of governance standards.

Incentives attract engagement and contribute toward the preservation of network integrity.

Model Deployment and Evaluation

The globally validated model has been distributed onto different nodes to perform inference or further enhancement.

Performance metrics and feedback gathered iteratively and successively enhance the model in successive federated rounds of learning.

With blockchain and FL, organizations could tap into the power of decentralized management of data along with the collaborative training of AI models, all while guaranteeing privacy and being transparent in safely executing operations.

6.6 CASE STUDIES AND APPLICATIONS OF IOT WITH BLOCKCHAIN-INTEGRATED FEDERATED LEARNING

Here are the detailed case studies and applications of BIFL for IoT-based smart applications in various domains [29]:

Case Study 1: Healthcare Monitoring

Problem Statement: Hospitals and healthcare providers are increasingly relying upon IoT devices for continuous patient monitoring. It generates a massive amount of very sensitive data such as vital signs, medical histories, and treatment outcomes.

Application: FL can enable the coordination of training of models by AI on different hospitals in an entirely private manner with respect to patient identity. Each hospital could still respect regulations on data privacy, for example, in the United States, under HIPAA, by maintaining its control over its data.

Blockchain technology strengthens this process by:

Data Integrity: Applying blockchain to the record of model updates and transactions will ensure that the data is not changed and reliable in the FL process.

Auditability: Blockchain will be applied to the record of model updates and transactions to ensure no data is changed in the FL process.

The cryptographic technology of the blockchain ensures secured transactions of data. Patients' records involved in this process are ensured to be of confidence and integrity.

Case Study 2: Smart Cities

Problem Statement: Cities worldwide use IoT sensors to monitor urban characteristics in the plan by using traffic flow, air quality, energy consumption, and even more metrics to optimize city operations, thereby enhancing citizen welfare.

Application: FL allows different city departments – including transportation, environment, and energy – to jointly train AI models on locally stored IoT source data.

It further enhances this process by integrating blockchain:

Data Integrity: Utilization of blockchain technology is going to ensure that data being sourced from the IoT will both be authentic and trustworthy, minimizing data interference or manipulation prospects.

Decentralized Governance. Smart contracts on blockchain ensure transparent governance of updates and then consensus mechanisms between the departments of a city.

Interoperability: Blockchain protocol allows for communication and secure transactions between heterogeneous IoT networks and municipal departments, which increases the dependability and efficiency of smart city applications as a whole.

Case Study 3: Industrial IoT (IIoT)

Problem Statement: Industry manufacturing leaders make use of IoT sensors in monitoring equipment performance, predicting maintenance needs, and optimizing production processes [30].

Application: FL can facilitate the collaborative work of multiple factories within a supply chain or industrial network in order to train AI models on their local IIoT data.

Blockchain integration provides the following:

Data Privacy: Blockchain guarantees that industrial-specific sensitive data, including production metrics and the status of equipment, will not be leaked during FL.

Transaction Security: Using the consensus mechanisms and cryptographic protocols in the blockchain will facilitate secure transactions and verifiable updates in different industrial sites.

Immutable Records: Decentralized ledger technology in blockchain ensures that a permanent record of model training processes and data transactions is maintained, which significantly enhances both transparency and auditability in the industrial IoT landscape.

Case Study 4: Financial Services

Problem Statement: Financial institutions are using IoT devices in support of helping ensure real-time transaction oversight, stronger fraud detection, and improvement of their processes on risk management [31].

Application: FL enables the banks and financial services providers to train AI models on a decentralized IoT data source with privacy and confidentiality of the data.

Blockchain enhances this process by:

Data Confidentiality: Use of blockchain ensures private and secure sensitive financial transaction data is maintained during FL.

Transaction Transparency: The transparent ledger in blockchain permits auditable traceable transactions across the financial systems. Model updates are ensured to be sound.

Regulatory Compliance: Blockchain happens to be in compliance with financial regulations, such as GDPR and PSD2, so that the practices followed in data handling and model training comply with legal and regulatory requirements, and thus, there is trust in and accountability toward the financial services sector.

Case Study 5: Agricultural Monitoring

Problem Statement: Farming companies and cooperatives use IoT sensors to measure soil quality, monitor crop growth, and measure all environmental factors, thereby improving agricultural productivity.

Application: FL allows agricultural stakeholders to train diverse IoT data sources in an AI model, all the while respecting sovereignty and personal privacy.

Blockchain integration provides:

Data Sovereignty: Blockchain will ensure individual farms have ownership and control over their agricultural data so that no third party can access them illegitimately or steal them.

Data Integrity: Through blockchain, authentication and reliability of the agricultural IoT data used in the model training process will eliminate the risk of manipulated data or fraud.

Supply Chain Transparency: The decentralized ledger system of blockchain facilitates transactable and traceable transactions in agricultural cooperatives while greatly helping in developing the sense of trust and accountability in the agricultural supply chain.

FL that incorporates blockchain technology provides seamless smart applications based on IoT for diverse domains by ensuring data privacy, integrity, and transparency. These technologies allow for collaborative training of AI models without violating the privacy of sensitive IoT data and, therefore, are suitable for applications where data security and trust will be relevant considerations.

6.7 PERFORMANCE EVALUATION AND METRICS

Blockchains-added federative IoT-based smart applications require performance evaluation and metrics because that is usually necessary for assessing and evaluating the effectiveness and efficiencies of the system [32–34]. A structured approach is essential for defining performance evaluation criteria and selecting appropriate metrics to accurately assess system effectiveness:

Training Efficiency Metrics

Round Trip Time: A time is defined as the time taken from the model update traveling from the edge device to the central server and back. Lower round trip time is a faster convergence of the model.

Communication Overhead: The amount of data that flows between the edge devices and the central server during any round of communication. The reduction of overheads leads to a reduction in both the consumption of bandwidth and latency.

Convergence Rate: It calculates the total number of iterations that the model using FL takes to achieve the accuracy level the user aspires for across all the devices participating within it.

Blockchain Integration Metrics

Transaction Throughput: Measure the number of transactions, or model updates, that are processed by a blockchain network within a period. The greater the throughput, the more scalable.

Latency: Determine the period it takes from the time a transaction (model update) enters into a blockchain until it is confirmed final. Minimal latency lets for speedy agreement and authentication.

Security Overhead: Evaluate the computing and storage overhead generated from a perspective of conducting blockchain operations. It should not severely impact the performance of FL.

Model Performance Metrics

Accuracy: The predictive accuracy of the model FL on edge devices. Compare before and after FL in accuracy metrics to look into improvement.

Robustness: Introduce into study the effectiveness of a model's maintaining performance when its conditions fluctuate, as well as when edge device capabilities vary. The IoT models in such cases that show robustness can perform effective generalizations in various IoT-based settings.

Privacy and Security Metrics

Privacy Preservation: It investigates the effectiveness of privacy-preserving methods, such as differential privacy and encryption, in protecting the sensitive information during FL as well as blockchain transactions.

Data Integrity: Verify the integrity of model updates and transactions stored on the blockchain to prevent tampering or unauthorized access.

Adversarial Robustness: Evaluate the resilience of the FL and blockchain integration against adversarial attacks and malicious participants.

Resource Utilization Metrics

Compute Resources: Evaluate the consumption of CPU and GPU on edge devices as well as the central server during phases of FL and blockchain operations.
Storage Utilization: Storage to hold blockchain data, logs for each transaction, and parameters of every model. Use optimization strategies to satisfactorily face the constraints on resources.

Operational Metrics

A storage system meant to archive blockchain information, logs related to every transaction, and specifications related to each model.
Fault Tolerance: Determine the efficiency of the system in terms of node failure and network failure without affecting the data consistency as well as correctness for the model.

User Experience Metrics

Usability: Feedback from the users involving developers, administrators, and even end-users on the convenience of implementing and managing BIFL in IoT applications contexts should be acquired.
Scalability: Assess the scale at which the system can accommodate increased numbers of edge devices and IoT endpoints so performance and efficiency are maintained.

Environmental Impact Metrics

Energy Efficiency: Measure the energy usage with which FL cycles and blockchain processes are related. That improves the protocols and algorithms to be used to minimize energy usage.
It provides a holistic framework for performance evaluation in terms of effectiveness, efficiency, and practicality achieved by the blockchain integration of FL for IoT-based smart applications. The monitoring of such metrics would play a crucial role for the improvement of system performance, security, and assurance to reliable operation in different practical deployment scenarios.

6.8 FUTURE DIRECTIONS AND RESEARCH CHALLENGES

Blockchain and FL are considered to be a promising combination in the future context of IoT-driven smart applications. However, they face several challenges that must be addressed before they can fully realize their potential. Possible future directions and challenges in these areas are outlined below..

6.8.1 Future Directions

Enhanced Privacy-Preserving Techniques:

Develop advanced FL frameworks capable of addressing the privacy of data safety issues; this should include decentralized setting, like IoT context. This deals with the study of strategies such as homomorphic encryption, differential privacy, and FL with secure aggregation.

Scalability and Efficiency Improvements

Scaly architecture and algorithms that support the large-scale, widespread deployment of IoT devices performing FL. Solutions must be proposed that improve communication efficiency, computation efficiency, and required storage efficiency while ensuring real-time response.

Interoperability and Standards

Those interoperability standards for the integration of the different blockchain platforms through FL framework shall include the necessity of ensuring compatibility among the various devices of IoT and blockchain networks for hassle-free data exchange and streamlined transaction management operations.

6.8.2 Research Challenges

Data Heterogeneity and Quality

In fact, since the IoT devices are distributed, they may differ significantly in terms of data format, reliability, and volume. Methods for effective preprocessing and quality assurance of data in the FL environment should be developed.

Blockchain Scalability and Performance

Scalability issues with blockchain networks, by which transaction throughput, latency, and storage capacity are likely to impact the operational efficiency of FL, need to be addressed; thus, blockchain performance enhancements should not degrade security guarantees.

Trust and Governance Issues

Address trust and governance challenges in decentralized FL ecosystems where multiple stakeholders (IoT device owners, data providers, model developers) interact. Define mechanisms for transparent decision-making, dispute resolution, and accountability.

Energy Efficiency and Resource Constraints

Optimize energy consumption and resource utilization in BIFL systems, considering the limited computational and battery resources of IoT devices. Develop energy-efficient algorithms and protocols for decentralized model training.

Real-World Deployment and Validation

Validate the feasibility and effectiveness of BIFL in real-world IoT applications across diverse sectors (healthcare, smart cities, agriculture). Conduct pilot studies and field trials to assess scalability, security, and user acceptance.

Education and Skill Development

Promote education and skill development in blockchain technology, FL, and IoT among researchers, developers, and practitioners. Foster interdisciplinary collaboration to tackle complex challenges and drive innovation in this emerging field.

Addressing these future directions and research challenges will open the door to leveraging the combined power of blockchain and FL for enhancing privacy, scalability, security, and efficiency in IoT-based smart applications and, as such, advance the state-of-the-art in decentralized machine learning paradigms.

6.9 CONCLUSION AND FUTURE WORK

In conclusion, the integration of blockchain with FL will greatly leapfrog the IoT-based smart applications in areas related to data privacy, security, and efficiency issues. The decentralized design of a blockchain framework will allow transparent management with tamper evidence, while FL designs facilitate collaborative models of training without any centralization of their data collection. Blockchain technology in conjunction with FL can build mutual trust and security within the IoT ecosystems and facilitate further sophistication of local AI models, tuned into a dataset of specific regions. BIFL thus represents an important leap forward in realizing a vast scope of possibilities seen with IoT technologies in different fields that include healthcare and smart urban development.

There are several directions for future research and development, mainly on applying BIFL to IoT applications. The first area that needs exploration is the optimization of the efficiencies with which FL algorithms compute and communicate in order to accommodate resource-constrained IoT devices. Another option would be for the development of strong incentives within blockchain protocols to encourage IoT devices' owners to participate and share data in FL.

REFERENCES

1. Lyu, L., Yu, H., Wu, X., Zhao, Q., & Xie, Q. Privacy-preserving blockchain and federated learning for IoT-based smart applications. *IEEE Network*, 34(3), 234–241, 2020.
2. Kang, J., Xiong, Z., Niyato, D., & Zhao, P. Incentive mechanism for reliable federated learning: A blockchain-based approach. *IEEE Internet of Things Journal*, 6(5), 10700–10712, 2019.
3. Yang, Q., Liu, Y., Chen, T., & Tong, Y. Federated machine learning: Concept and applications. *ACM Transactions on Intelligent Systems and Technology* (TIST), 10(2), 1–19, 2019.
4. Zhang, W., Lu, R., & Cheng, N. Blockchain-integrated federated learning for smart healthcare applications in IoT. *IEEE Internet of Things Journal*, 8(5), 3451–3465, 2021.
5. Li, T., Sahu, A. K., Talwalkar, A., & Smith, V. Federated learning: Challenges, methods, and future directions. *IEEE Signal Processing Magazine*, 37(3), 50–60, 2020.
6. Lu, Y., Huang, X., Dai, Y., Maharjan, S., & Zhang, Y. Blockchain and federated learning for privacy-preserved data sharing in industrial IoT. *IEEE Transactions on Industrial Informatics*, 16(6), 4177–4186, 2020.

7. Gao, Y., Yu, W., Liang, H., & Tang, J. Blockchain-enabled IoT-based smart applications. *IEEE Communications Magazine*, 58(3), 89–95, 2020.
8. Kang, J., Yu, R., Huang, X., Maharjan, S., & Zhang, Y. Enabling localized peer-to-peer electricity trading among plug-in hybrid electric vehicles using blockchain and federated learning. *IEEE Transactions on Industrial Informatics*, 14(10), 4624–4634, 2018.
9. Zhang, K., Ni, J., Yang, K., Liang, X., & Shen, X. S. Federated learning for smart city applications. *IEEE Communications Magazine*, 58(6), 42–47, 2020.
10. Wang, H., Kaplan, L., Han, Y., & Chai, S. Blockchain for decentralized federated learning in IoT. *IEEE Internet of Things Journal*, 6(5), 7875–7884, 2019.
11. Chen, X., Hu, W., Sun, S., & Chen, Y. Blockchain-enabled federated learning for autonomous driving in vehicular networks. *IEEE Network*, 35(3), 185–191, 2021.
12. Nguyen, D. C., Ding, M., Pathirana, P. N., & Seneviratne, A. Federated learning for COVID-19 detection with blockchain-based security in IoT networks. *IEEE Internet of Things Journal*, 8(21), 17368–17379, 2021.
13. Zhou, Z., Wu, H., Sun, H., & Xu, Y. Blockchain and federated learning for decentralized energy trading and management in smart grids. *IEEE Transactions on Smart Grid*, 11(3), 1678–1688, 2020.
14. Sun, H., Li, Z., & Xu, X. A blockchain-enhanced federated learning framework for mobile edge computing in IoT environments. *IEEE Network*, 34(3), 35–41, 2020.
15. Tang, J., Xie, Z., Zhang, Y., & Liu, H. Blockchain-based federated learning and security for edge-enabled IoT. *IEEE Communications Magazine*, 57(7), 12–17, 2019.
16. Peng, S., Wang, Q., Zhang, X., & Yang, L. Blockchain-powered federated learning for IoT-based smart grid security. *IEEE Transactions on Industrial Informatics*, 17(5), 3489–3496, 2021.
17. Prasanth, A., & Jayachitra, S., A novel multi-objective optimization strategy for enhancing quality of service in IoT enabled WSN applications. *Peer-to- Peer Networking and Applications*, 13, 1905–1920, 2020.
18. Hao, Z., Wu, L., & Wang, X. Distributed federated learning with blockchain technology for IoT data privacy. *Future Generation Computer Systems*, 112, 127–134, 2020.
19. Lathika, V., & Jayachitra, S., Secure multi-party computation for collaborative data analysis in network security. *International Conference on Intelligent and Innovative Technologies in Computing, Electrical and Electronics (IITCEE)*, 1–5, 2024. doi 10.1109/IITCEE59897.2024.10467913.
20. Gunduz, D., Suseelam, B., & Lim, Y. Blockchain-based federated learning for IoT privacy. *IEEE Transactions on Neural Networks and Learning Systems*, 31(4), 978–985, 2020.
21. Liu, C., Ma, X., Xu, X., & Wang, P. A blockchain-based federated learning framework for real-time applications. *IEEE Access*, 7, 68413–68423, 2019.
22. Kumar, G., & Aravindh, R. Seismic Image Pattern Analysis Using Fuzzy Logic Controller, *3rd International Conference on Intelligent Sustainable Systems (ICISS)*, Thoothukudi, India, 998–1002, 2020.
23. Wan, S., Li, W., & Wang, J. Blockchain-based federated learning framework for smart manufacturing in IoT. *IEEE Internet of Things Journal*, 8(9), 6737–6745, 2021.
24. Sudha, M., & Mahesh Kumar Reddy, V. Optimizing intrusion detection systems using parallel metric learning. *Computers and Electrical Engineering*, 110, 1–15, 2023.
25. Al-Rakhami, M., & Siddiqui, S. Secure and privacy-preserving federated learning for IoT using blockchain. *Future Generation Computer Systems*, 115, 157–166, 2021.
26. Sharma, P. K., Moon, S. Y., & Park, J. H. Blockchain for decentralized IoT: Enhancing federated learning for security. *IEEE Communications Magazine*, 57(7), 37–43, 2019.

27. Nigam, C., & Kirubasri, G., Real-time object detection on edge devices using mobile neural networks. *International Conference on Intelligent and Innovative Technologies in Computing, Electrical and Electronics (IITCEE)*, 13, 1–4, 2024.
28. Zhang, Y., Yang, J., & Tang, S. Blockchain-enhanced federated learning for security in smart cities. *IEEE Transactions on Cybernetics*, 50(4), 1437–1446, 2020.
29. Qiu, Y., Wang, L., & Feng, L. Federated learning and blockchain-based security framework for IoT applications. *Journal of Network and Computer Applications*, 115, 122–130, 2021.
30. Xiao, L., Xu, Z., & Wang, X. Blockchain-federated learning and IoT: Emerging applications in autonomous vehicles. *IEEE Network*, 34(3), 36–41, 2020.
31. Haque, M. A., & Akhtar, S. Blockchain-federated learning for IoT in smart agriculture. *IEEE Access*, 8, 152123–152133, 2020.
32. Han, J., Park, J. H., & Kim, Y. Blockchain-based federated learning framework for IoT devices. *IEEE Transactions on Green Communications and Networking*, 5(1), 109–118, 2021.
33. Dai, H., Zhang, Q., & Chen, L. Federated learning and blockchain for cybersecurity in IoT networks. *IEEE Communications Surveys & Tutorials*, 22(3), 2033–2045, 2020.
34. Ren, Z., Yang, F., & Zhou, G. Blockchain-based privacy-preserving federated learning in IoT networks. *IEEE Network*, 35(3), 34–40, 2021.

7 Federated Learning in Heterogeneous Unmanned Aerial Vehicle

N. Sathish, D. Manibharathi, V. Yokesh, and Danilo Pelusi

7.1 INTRODUCTION

Federated learning (FL) is a relatively new concept, specifically in machine learning, that discusses the idea of different devices working together to train a pertinent model without sharing personal data. FL is most advantageous where utmost consideration for data privacy is a must, as FL ensures data are contained in each device used [1]. The clients perform local model training on data available within their systems and only periodically send learnings in terms of a change from the original model, such as gradients, to a central server [2]. Afterwards, the server integrates these changes to make an overall global model, which will return to the clients that took part in the learning process. It makes it easier to train models for dispersed and divergent systems using various and remote datasets and avoid some privacy-related issues that normally come hand in hand with such processes. Further, it does not require a large exchange of data and therefore does not require huge costs and resources [3].

FL is most valuable in use cases such as smartphones, IoT devices, and unmanned aerial vehicles (UAVs), where large amounts and various types of data are generated [4, 5]. FL is a great solution for solving problems arising from data heterogeneity, privacy, and limited communication in distributed networks. Its strength is that it is able to store local data and at the same time it is making a contribution to the global model. The general structure of FL is shown in Figure 7.1. Therefore, FL is central toward enhancing the effectiveness and security of UAV systems. In these networks, UAVs often work in diverse and distributed scenarios and gather enormous volumes of data.

Classical methods of machine learning require data to be stored normally in some central location for training, which has serious issues with data security [6]. FL allows UAVs to jointly learn on the same model while keeping disparate data decentralized. This not only ensures that privacy is kept but also reduces the need for huge communication bandwidth to some extent, as only those updated to the model need to be communicated [7]. Besides, FL improves the adaptability and intellectual capacity of UAV networks by allowing the acquisition and application of immediate insight and

DOI: 10.1201/9781003532323-7

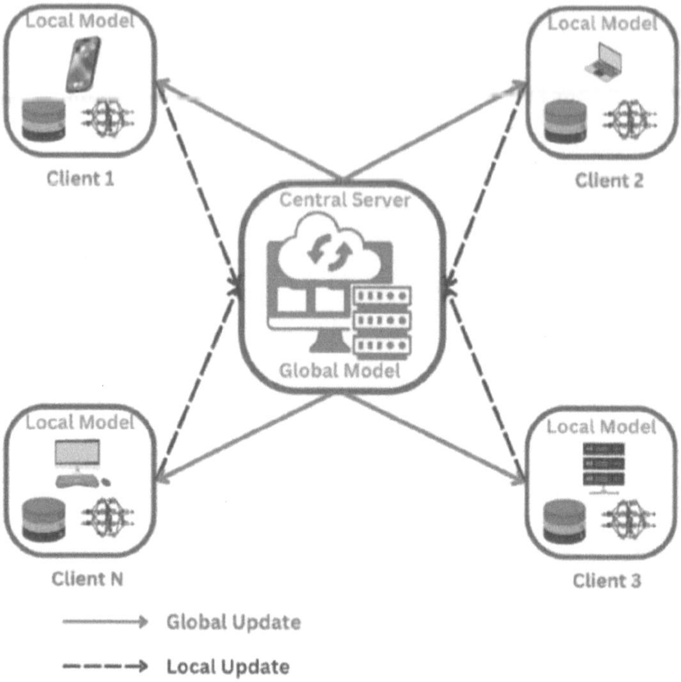

FIGURE 7.1 Architecture of FL.

intelligence from local data to make good decisions and carry out better operations in a given loophole. As the application of UAVs increases, mainly in surveillance, delivering goods, and handling disasters, FL is becoming more important. To ensure safe, efficient, and scalable learning, FL is required at the varying networks across the implementations [8].

7.2 BACKGROUND AND RELATED WORK

The proliferation of UAVs in many areas increases the need for efficient signal-processing approaches that address different challenges, particularly privacy, message transmission, and system enlargement. FL has been one of the significant approaches that have enabled various UAVs to work together in training other machine learning models with minimal data centralization and security threats. This segment provides information on the core of FL, its application in the UAV networks, and a detailed analysis of some relevant research to illustrate its benefits and existing innovations. From these facets, the goal is to provide a comprehensive understanding of how FL addresses essential issues in UAV networks and its prospects for future advancement.

7.2.1 Basics of UAV

Drones are more formally referred to as UAVs, and this technology is relatively young, but it has a very versatile application in many industries. UAVs can be aircraft

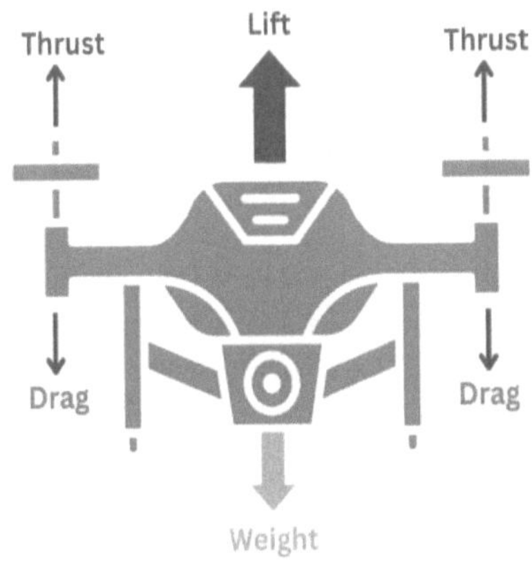

FIGURE 7.2 Drone anatomy.

systems or flying vehicles that can operate autonomously or remotely without a pilot [9]. Ground operators or self-controllers with sophisticated onboard systems or pre-coded flight sequences can operate the vehicles. They range in size from tiny quadcopters used for recreational aerial photography to enormous, complicated military aircraft used for battle and surveillance. Small UAVs are powered by electric motors and propellers, and their basic components are the body or structure, engine and propeller systems, avionics (sensors, cameras, GPS, and other communication modules), and a system for relaying signals to the operator or control center [10]. UAVs can carry commodities, take high-resolution photos, and sample the environment when these components are integrated.

Drone anatomy is essential to understanding UAV flight dynamics, control systems, and legal requirements, and it is illustrated in Figure 7.2. Flight dynamics encompass lift, thrust, drag, and gravity. Vehicle stability, controllability, altitude, and direction are affected by these factors. Manually or automatically control UAV dynamics to maneuver or follow flight patterns. Software reads sensor data from accelerometers and gyroscopes to change UAV location and velocity [11]. Legal frameworks for operator certification, no-fly zones, and maximum altitude vary by country but are important [12]. If human aircraft are present, these standards should assure safe flight. UAV operators, developers, and legislators can use these notions as technology advances.

7.2.2 FL Principles

FL is based on three main principles, namely decentralization, the protection of individuals' rights to privacy, and the improvement of the cooperative model. Basically, FL prevents data from moving to a central location but remains with the devices, a process known as federated training. One of the benefits of this approach is that it solves privacy concerns and reduces the probability of unauthorized access to data, as the data remains within the company's network boundaries. Another important element is safe aggregation, in which only deltas, such as gradients, are transmitted to a central server. Such updates can often occur in encrypted forms so that even if the data is intercepted, it cannot be easily decoded.

Every one of these changes is then combined by the central server to enhance the global model before it is relayed back to the devices. When the model does not reach the prescribed level of accuracy then it has to be iterated this way until the right level of accuracy is achieved. Another aim of FL is to minimize the communication load by increasing the proportion and frequency of the updates that are sent from the device to the server. Applying these rules, FL is an effective procedure that does not disclose personal information and strengthens machine learning models.

7.2.3 Review of Existing Work on FL in UAVs

Sattler et al. proposed a communication-efficient FL method that utilizes gradient sparsification and quantization to decrease the amount of communicated updates. This is especially advantageous for UAVs with limited communication capabilities [13]. Wang et al. put forward an adaptive FL approach that can dynamically adjust the frequency of the model update based on the network condition and the capability of UAVs. It allows to the minimization of such trade-offs between model accuracy and communication costs when using prototypes [14]. Chen et al. presented an FL system for UAV networks in which model pruning and quantization are used to reduce computational load. This makes it possible for UAVs to train in order to improve their operation without expending their resources [15].

In an FL method named "hierarchical FL", Zhao et al. proposed that UAVs with better computational performance work as local aggregators to reduce the substantial burden on other UAVs with inferior processing capacity and improve the efficiency of the learning process [16]. To overcome the problem of diversity of data, Li et al. suggested the FL method that contains a personalized learning part. This makes it possible for each of the UAVs to correct the probability distribution given in the global model with regard to its local dataset distribution [17].

In their work, Ghosh et al. found that they proposed a multitask FL framework whereby each UAV trains its model suited to respective tasks, yet the UAVs all learn from a single representation. It also helps prevent issues with different data sources in the analysis process of a given subject [18]. Yang et al. introduced a robust aggregation mechanism for FL, which guarantees the privacy of each UAV update, even when there are malicious nodes present [19]. Bonawitz et al. devised a method for FL that ensures privacy by employing differential privacy approaches. This method safeguards the confidential data of UAVs while training the model [20].

7.3 SYSTEM ARCHITECTURE

7.3.1 UAV Network Topology

UAV network topology is concerned with how UAVs are positioned in a network and how these aerial entities perform as well as how they communicate with one another to fulfill their intended purpose. The relevance and importance of understanding how to configure a UAV network is needed since it is the central factor in determining the capacity of the network in terms of productivity, reliability, and scalability of the network in its operations. As UAV operations involve dynamic scenarios, this also means that the network architecture designed for UAV networks must be able to cater for the mobility, distance, and connectivity that are prone in UAV networks [21].

UAV networks commonly employ many network topologies, each possessing distinct merits and drawbacks. The different topologies are star topology, mesh topology, tree or hierarchical topology, and hybrid topology.

7.3.1.1 Star Topology

All UAVs are connected by a direct link while all UAVs are connected to a central node that can be a ground control station (GCS) or a highly capable UAV. Thus, the architecture in question provides the necessary management of communications since all data are processed through the central point. But with the star topology, there is only one point of failure and this makes the topology vulnerable. If the central node is compromised through a malicious attack or loss of link, maybe the whole network system would be dysfunctional. Despite the protection weakness of star topology, it is useful where centralized command and control are essential, and the UAVs are positioned closely to the central node.

7.3.1.2 Mesh Topology

Mesh topology is another topology type and is distinguished by the fact that each node has the possibility to communicate with a lot of other nodes directly. This configuration creates a strong network of data transmission having multiple paths, hence increasing the rate of failure and duplication. Data can then route to other paths if the UAV becomes disconnected or a communication connection fails, ensuring that the network is always strong. As such, described mesh networks are most helpful where multiple UAVs are required to survey vast areas or operate in complex and dynamically changing environments. Nonetheless, it may require more complicated networking protocols and may also require more processing power to implement routing techniques to maintain the strength of the mesh topology [22].

7.3.1.3 Tree or Hierarchical Topology

The tree or hierarchical topology means the UAVs are grouped in clusters or layers. Each cluster has a node that gathers information from the constituent nodes and communicates with the superior nodes or ground station nodes. Like the star topology, certain UAVs in this topology have a heavy communication burden while other UAVs have low or no burden at all; like the mesh topology some UAVs in this topology have a communication redundancy while others have no redundancy at all. Hierarchical topologies have the capabilities for scalability and comprehensive UAV management and therefore are suitable for coverage tasks like surveillance, search and rescue, and environmental studies.

7.3.1.4 Hybrid Topology

The hybrid topology combines components from the earlier stated topologies to benefit from their advantages with minimal vices. For instance, a hybrid network may use star topology for connection between the clusters of UAVs that are close to a base station while the use of a mesh topology for the connection between different clusters of UAVs. This makes it possible to have changes in the network design and the performance and reliability of communication, depending on the mission goals and objectives.

7.3.2 FL Framework for UAVs

A UAV FL framework allows several UAVs to train a single machine learning model [23]. It protects data and reduces communication. Traditional machine learning methods collect data from multiple sources into a single server or data center for model training. This method has substantial privacy problems and consumes a lot of communications resources, especially in UAVs, where bandwidth is limited and data transfer is expensive or unavailable. FL addresses such issues by decentralizing data with each UAV. Each UAV builds its local model from data it collects and only shares model updates (such as gradients or weights) with a central aggregator, such as a more advanced UAV, edge server, or ground station. In distributed systems, the central aggregator collects updates from all UAV partners and merges them to update the global model.

The updated global model is given to UAVs, which continue local learning. This procedure continues until the global model is accurate enough for fresh datasets. UAVs do not exchange their raw data; therefore, they are less likely to be hacked and meet data protection regulations. FL also manages communication costs better than other methods since it transmits only model descent instead of raw data. However, non-IID (non-independent and identically distributed) data makes FL integration in UAV networks difficult. This is because UAV data might vary greatly depending on its location, purpose, and sensors. Different forms of data can affect model performance or stability. Because of this, machine learning algorithms need to be made to handle a variety of data distributions.

UAV networks are highly dynamic, with continually changing connection and network properties owing to UAV movement, requiring strong and dynamic communication protocols. Another key issue with UAVs is their limited energy capacity, which hinders computationally intensive activities and long-haul communications. FL frameworks for UAVs use model pruning and quantization to reduce model update size to conserve power. Decentralized aggregation allows direct UAV aggregation without a master aggregator, making the network more adaptable and scalable. These frameworks protect UAV data with differential privacy and secure multiparty computing. To ensure this, Modest suggests protecting actual data below model change messages. UAV FL frameworks can provide great efficiency, security, and intelligent collaborative learning in unstable circumstances. It could lead to more autonomous and efficient UAVs [24].

7.4 FL PROCESS IN UAVS

FL in UAVs is a distributed machine learning technique that enables several UAVs to collectively train a common global model without exchanging their raw, locally

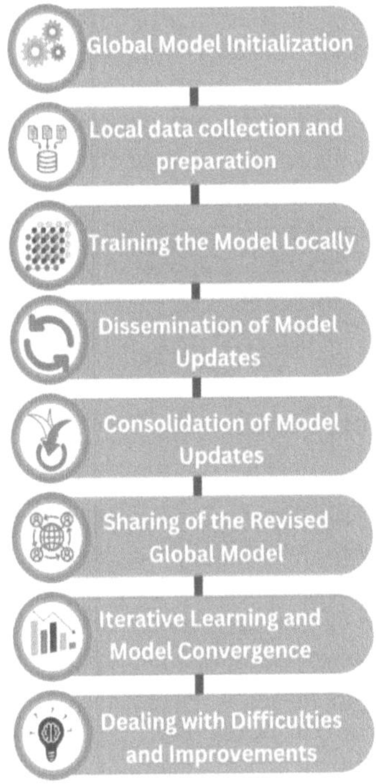

FIGURE 7.3 FL process in UAV.

gathered data. This approach is especially beneficial for UAVs, which frequently function in settings where the protection of data, the safeguarding of information, and the optimization of communication are of utmost importance. The FL process in UAVs is organized into many crucial stages, each specifically intended to guarantee effective and protected knowledge acquisition over a decentralized network [25]. The step-by-step process is demonstrated in Figure 7.3.

7.4.1 Global Model Initialization

The FL process begins by initializing a global model that can be kept on a central server or any base station, including GCS, edge server, or any selected leader UAV having better computation power. A global model essentially implies the architecture of an artificial neural network designed to perform specific tasks like object recognition, sensing the surroundings, or finding abnormalities. Usually, the model parameters (weights) are initialized at random; however, if the algorithm can function with some existing data, other techniques may be employed.

7.4.2 Local Data Collection and Preparation

Every UAV operates independently in the designated area or task area and always collects data, including through the use of incorporated sensors, cameras, or other tools. Such information may include photographs, videos, sensor information, and telemetry data, depending on the UAV mission objectives. Before feeding data to the machine learning model, unprocessed data is locally preprocessed on each of the UAVs to make it in the right format for training. The preprocessing stages that are involved in the preparation of the data for analysis may include normalization, augmentation, and filtering or feature extraction. More of the fact that the general steps depend on the data and the learning job to be done. Most importantly, in this stage, the extracted information is stored locally in each UAVs memory and is not uploaded to a central server to reduce data disclosure and communication burden.

7.4.3 Training the Model Locally

After data preprocessing, all the UAVs locally perform the training on the model they initialized using their collected data. What this process involves is that several rounds (also known as epochs) of training are performed, and in this process the model learns how to make predictions or classifications depending on the data it has been trained on. The training phase involves techniques such as stochastic gradient descent to adjust the weights of the model and hence minimize the loss function, which measures the difference between the model predictions and the actual data. As each UAV uses data it gathers locally to train its model, the process of learning is highly specialized and tailored to the environment and conditions that the UAV operates in. This leads to updating the model to represent the innate variability of data that are distributed across multiple UAVs.

7.4.4 Dissemination of Model Updates

The local training phase ends with every UAV having a set of model updates, typically the gradients or weight differences computed during the local model training. Here, the updates, rather than the raw data, are transferred to a central aggregation server or a leader UAV that has been designated for the purpose. Maintaining that sensitive or private data stays on the UAV is important for privacy because it minimizes the risk of leaks or breaches of privacy. Coordination of the communication phase must be done efficiently to ensure that batch bandwidth and batch time are optimized and delay costs are cut in situations where connectivity may not be reliable. Techniques like model pruning, quantization, and activation mechanisms for pruning and reducing the amount of data to be sent are some techniques that are used to reduce the size of the model updates that are sent and, hence, make the communication process more efficient.

7.4.5 Consolidation of Model Updates

The main server or the leader UAV receives the updates to the model from all the UAVs involved and brings them together to update the global model. Some of the popular techniques for aggregation include federated averaging (FedAvg), in which the server makes a weighted average of the updates from the different UAVs taking into consideration the amount of data samples they used in local training. This method results in an all-new global model parameter that has been learnt from all the UAVs. In addition, aggregation can contain more measures for its security and reliability, like the use of secure aggregation, which does not let one party learn about the updates of the other party's UAVs or the use of outlier detection so as to reduce the quantum of bad or even wicked UAV in the UAVs network.

7.4.6 Sharing of the Revised Global Model

Once the global model is updated, new model parameters are broadcast to all UAVs in the network. After that, each UAV updates the latest global parameters to its specific model, which hence brings all the UAVs in line with the present state of the global model. This stage ensures that each of the UAVs benefits from collaborative learning even though a particular UAV may only have access to limited local data. Other sets of local training may then be employed to fine-tune the new model, hence leading to improvement of the model and its performance across the network.

7.4.7 Iterative Learning and Model Convergence

FL is an iterative process. Steps 3 through 6 are repeated several times. Each cycle involves local training, model change exchange, aggregation, and broadcasting of the modified model. This iterative process persists until the global model converges, meaning that additional training rounds don't improve the accuracy or performance of the model. Convergence is usually measured by some defined criteria, such as the maximum number of communication cycles, a threshold to the loss function, or some kind of performance objective. This makes FL an iterative process, where it empowers a model to improve its performance incrementally by capturing diverse and decentralized knowledge from all UAVs for a more robust and universally workable model.

7.4.8 Overcoming Challenges and Improvements

FL processes in UAV networks face several inherent challenges, including managing non-IID data, lowering communication costs, ensuring model convergence, and maintaining privacy and secrecy. To mitigate these issues, a lot of improvements have been adopted, one of which includes personalized FL, where each UAV is allowed to maintain a model tailored toward its underlying distribution while contributing to a global model. Differential privacy can be applied to randomly perturb the model updates, making it impossible to infer specific data points even if intercepted. Other communication solutions, such as federated dropout, reduce communication costs, thereby increasing scalability by running only a subset of the UAVs in each cycle. To this end, such robust aggregation techniques would diminish the impact of untrustworthy or even malignant UAVs, ensuring integrity and dependability in the FL procedure.

FL is a very strong approach that allows everyone in collaborative machine learning in decentralized ways preserving everyone's data but having efficient communications. It operates on local data, synchronizes updates in a secure manner, and fine-tunes the relative model in creating high-value models to be deployed in varied environments of UAV networks. Despite these challenges, the FL approach in UAVs can still be made feasible and effective for many purposes due to the ever-advancing FL algorithms, communication technology, and security.

7.5 CHALLENGES AND SOLUTIONS

FL is a possible mechanism to update the semantic communication (SC) model in a decentralized approach where no raw data should be transferred to the central server. The decentralized method benefited UAVs mostly because they often execute in environments where it is vital to preserve data confidentiality, coordinate scarce bandwidth, and where communication impediments play a massive role. However, one of the major challenges that arise with the integration of FL in heterogeneous UAV networks is as follows. The challenges are in factors such as the variation in the capability of devices, the network, the communication of data, and the constraints in energy [26].

7.5.1 Challenges

7.5.1.1 Hardware Heterogeneity

UAVs in a network usually are hardware diverse, meaning the differences in their processor power, memory capacity, and sensor types. These inconsistencies could lead to UAVs of different capacities. Some of the UAVs may be able to perform data analysis, model development, and training quicker than others. This version incorporates a bottleneck as the FL process relies on synchronous updates, which may be interrupted by slower devices thus slowing down the overall training process.

7.5.1.2 Unreliable Network Connectivity

UAVs operate in constantly changing environments where their capability of staying connected to a central server or talking to each other might be uncertain or temporary due to many factors like distance, signal blockage, and interference. System instability can cause postponed communication updates, thus leading to infrequent synchronization and the velocity of convergence as well as the general performance of the FL model.

7.5.1.3 Non-IID Data Distribution

The data collected by UAVs in different places may not be independent and identically distributed. For instance, UAVs conducting surveys in different regions may gather data with large variations in their characteristics and spatial arrangement. The data distribution variance may easily become a factor that would affect the training process. Models that were trained with non-IID data may converge at poorer solutions, which could result in the decrease of performance and generalization capability of the model.

7.5.1.4 Energy Limitations

UAVs have a limited battery capacity, and doing computational work, such as training machine learning models on-site, can drain their energy resources substantially. This constraint becomes the major problem for tasks needing the use of large amounts of energy, such as training deep learning models, as it provides UAVs only a limited number of training iterations before it becomes necessary to be recharged.

7.5.2 Solutions

7.5.2.1 Adaptive FL Algorithms

Perhaps some type of adaptive FL techniques can be used to account for the variations in hardware among UAVs. Algorithms could vary the computational load according to the abilities of each UAV. This allows more capable UAVs to perform more strenuous computations, while weaker devices are only burdened with lesser tasks. This strategy has the potential to mitigate the latency produced by slower devices and improve the overall efficiency of training.

7.5.2.2 Asynchronous FL

There is still a possibility to use asynchronous FL in the case of an unstable network connection. Unlike with the FL in which all the participating devices have to send their updates at the same time, the asynchronous methods enable UAVs to send their models at any given time depending on their status of connection. This is good since the number of UAVs continues to be trained at every moment and not all UAVs are dropped entirely at one point in time.

7.5.2.3 Data Augmentation and Regularization Techniques

To alleviate the non-IID-ness of the data, data augmentation methods can be employed in an attempt to artificially augment the diversity of the data collected by each individual UAV. However, by using regularization techniques like parameters or averaging with associated federated, it is possible to stop the global model from excessively paying attention to data of a few UAVs. This in a way improves the generalization capability of the model or the capacity to make a good prediction of the result given other conditions.

7.5.2.4 Energy-Aware FL

FL brings protocols that are in the development of energy-conscious FL protocols to increase the efficiency of training to conserve energy [27]. For example, UAVs can adjust the frequency of local updates using simplified models for the same or stop the process early to reduce workload. Last of all, FL tasks could be planned according to the state of charge of each UAV in order for no critical operations to become out of power.

7.5.2.5 Efficient Communication Strategies

To reduce the amount of information exchanged the models that are exchanged between the UAVs and the central server can be compressed using techniques involving quantization and sparsification. These methods lower the amount of information shared, hence reducing the use of bandwidth and energy. In addition, FL protocols could be created to only sporadically send the global model update or when there is a significant update to it, which takes a lot of communication to send.

Thus, it can be stated that FL may be successfully applied in a variety of different UAV networks if specific solutions to these problems can be found. Thus, this implementation enables decentralized, trusted, and privacy-preserving machine learning in shifting environments. Some of them can enhance the UAV network's capabilities in scenarios like surveillance, environment monitoring, and disaster relief, which require distributed knowledge and fast learning [28].

7.6 APPLICATIONS OF FL IN UAVS

UAVs with A2C capabilities can be created to enhance drone performance by learning from a swarm of drones without needing the raw data, as was shown in FL. This privacy-preserving approach is specifically beneficial when addressing the data containing sensitive information, or where the available network bandwidth is restrained [29]. The following are a few noteworthy use cases of FL in UAVs.

7.6.1 Surveillance and Security Monitoring

FL can be applied to UAVs deployed in the surveillance and security sector for object detection and recognition of suspicious activities. It will be appreciated if each UAV can collect some data about the environment of this or that territory, video image, or sensor data. For this purpose, this information is then used to construct a generalized world model that increases in efficacy over time. This arrangement makes tracking easier while at the same time eliminating the problem of compiling and sending information through other possible unsecure networks, hence enhancing the privacy and security of information like video.

7.6.2 Emergency Response and Administration

The major application of UAVs is in disaster situations where they can be employed in search and rescue operations, evaluating the degree of damage and logistical support. FL will allow the UAVs to learn cooperatively and develop their skills in search for individuals, evaluation of the extent of destruction, and identification of safe landing spots. They can explore different ways to improve the operational efficiency of UAVs in bandwidth-constrained or disconnected network conditions, particularly during disaster scenarios. Instead of transmitting the actual data, they can involve in transferring model updates.

7.6.3 Precision Agriculture

Precision agriculture, on the other hand, can be defined as the more accurate application of inputs through managerial technology and precise knowledge. This is due to the fact that UAVs are commonly being used in agriculture to evaluate the condition of crops, test the condition of the soil, as well as identify the presence of pests. FL also helps multiple UAVs located in different farms or regions to build improved and more reliable models that can be used in agriculture. For instance, UAVs that are working in a cluster may employ FL or machine learning approaches to boost up algorithms that identify the diseases. This could be done by training the algorithms on many different datasets from many different environments soils and crop varieties. Thus, the models are not only accurate but also more applicable, enabling their utilization in several agricultural scenarios.

7.6.4 Monitoring the Environment

Sensors mounted on UAVs are capable of obtaining invaluable environmental information relating to air quality, temperature, humidity, and level of pollution in the atmosphere. FL enables these UAVs to jointly build prediction models for environmental changes and the likely dangers, including fire outbreaks or floods. FL shields ecological information and minimizes the risk of compromising and unauthorized access as FL enables the stakeholders to accumulate knowledge without accumulating data to a central point.

7.6.5 Intelligent Urban Governance

They are applied for use in intelligent urban areas for traffic monitoring, infrastructure monitoring, and security. By using FL, these UAVs are enabled to learn about the surroundings and adapt their behavior to the constantly changing urban environment. For instance, UAVs may jointly enhance traffic prediction, designate optimal flight paths that do not pass through busy areas, and enhance image recognition algorithms to identify car crashes or problems with roads and bridges. This could lead to very smart and very flexible urban control systems that at the same time protect the privacy of individuals.

7.6.6 Optimization of Delivery and Logistics

UAVs are under investigation for their use in delivering services, and more specifically, the delivery of goods over short distances in urban and rural environments. FL allows the UAVs to teach each other about their experiences in different types of terrain, weather conditions, and city layouts and thus improves the optimization of delivery routes. It could also allow for better delivery efficiency and accuracy on behalf of the logistics companies, not to mention the protection of the customer's privacy and the logistics companies' compliance with data requirements by only passing along the model changes.

7.6.7 Monitoring and Preserving Wildlife

UAVs are used in conservation projects to monitor wildlife populations and track their movements and poaching. Using FL, UAVs can build robust models for species identification, behavior tracking, and threat detection without having to send large amounts of sensitive data (e.g., pictures/videos of endangered species), which could be used improperly. This way helps to keep the anonymity of the collected data and decreases the probability of revealing sensitive information about animal habitats.

7.6.8 Cooperative Cartography and Investigation

The UAVs that have cameras and LiDAR sensors may operate in parallel to create accurate models and maps of the surroundings. Such kinds of maps and models are extremely useful in such practices as surveys of terrains, archaeological explorations, and going to unknown lands. FL enables UAVs to obtain a higher resolution of the terrain by enabling them to alter their navigation and search procedures depending on new data received from other units, although this does not necessarily imply the existence of a central database. This method is particularly useful in areas of limited linkages or where the cost of transmitting data is an issue.

7.6.9 Self-guided Movement and Prevention of Collisions

Considering the usage of FL in UAVs, it can be concluded that FL can be beneficial for the improvement of a UAV's autonomous flight and collision avoidance program, particularly in a situation when these UAVs are flying in pairs or within a specific environment. The UAVs, through dissemination of learned information in the form of model updates to many different environments and flight conditions, would be able to learn more perfected and safer avoidance patterns, which would in turn reduce on probabilities of collision and enhance overall flight safety.

7.6.10 Crowd Sensing and Data Collection

UAVs can be utilized for crowd-sensing purposes to acquire data on population density, public gatherings, or emergency circumstances. FL enables UAVs to build models for crowd density estimation and crowd movement prediction using only local data, without compromising individual privacy. This program is really helpful (especially for urban planning, event management, and emergency response coordination).

UAVs can greatly improve their performance by employing FL and so operate in a much more intelligent, cooperative, and secure fashion across many applications. This approach enhances the utility of UAV networks while at the same time addressing problems of data privacy, communication bandwidth, and computational capability.

7.7 FUTURE DIRECTIONS

FL has shown great promise in enabling distributed and collaborative model training across UAV networks without the exchange of private information. The future of FL in UAVs may take many paths as the field matures and many of these paths could potentially increase the effectiveness of FL in a variety of UAV applications by overcoming existing hurdles and leveraging new technology. These include the creation of flexible algorithms that can work with various UAV abilities, the use of federated transfer learning to conserve usage, and the development of energy-efficient FL techniques to preserve battery life. Furthermore, through the incorporation of advanced privacy-preserving methods, the study of truly distributed FL approaches, and the use of federated reinforcement learning for real-time decision-making, the robustness and adaptability of UAV networks can be made even greater.

Further work will need to explore more sophisticated model combination approaches to better handle non-IID data and to capitalize on both edge computing and cloud computing to strike a balance between the tradeoff of decentralization and computational power. Real-time FL frameworks would help with immediate data processing, which is crucial for things like surveillance and disaster relief. However, it is imperative to work on interdisciplinary collaboration and rigorous testing of FL algorithms in various operational scenarios to make them perform better and be able to cater to the real problems in UAV networks. This will allow the networks to operate more independently and securely in a variety of applications. So, if FL follows these suggested routes, the FL area of UAVs will be on its way to better, more efficient, privacy-aware systems.

7.8 CONCLUSION

FL is one of the new promising methods that enables collaborative intelligence over heterogeneous UAV networks without sacrificing data privacy and security. FL allows UAVs to train models locally and send only the model updates, thus eliminating the need for centralized data storage, saving bandwidth, and improving data security. There are many challenges to implementing FL on UAVs with hardware heterogeneity, intermittent network connectivity, non-IID data distributions, and energy constraints. Nonetheless, the continuing changes in adaptive algorithms, the privacy-preserving approach, energy-efficient method, and decentralized architecture can be potential solutions to these problems. Thus, the use of formation flying in heterogeneous UAVs in the future is to improve the effectiveness of these strategies through the incorporation of new findings from other areas of study as well as through experimental validation.

FL will allow UAV networks to be more self-sufficient, productive, and secure as the fields of these technologies advance. It will, therefore, foster the progress of many industries such as security surveillance, in cases of disasters, agriculture, and environmental monitoring. In the end, FL would enable UAVs to coordinate their operations along with each other, which would alter the dynamics of operation in various fields.

REFERENCES

1. Fu, M., Shi, Y., Zhou, Y., Federated learning via unmanned aerial vehicle, *IEEE Transactions on Wireless Communications*, 23(4), 2884–2900, 2024.
2. Liu, X., Deng, Y., Mahmoodi, T., Wireless distributed learning: A new hybrid split and federated learning approach, *IEEE Transactions on Wireless Communications*, 22(4), 2650–2665, 2023.
3. Afzal, U. M., Abdellatif, A. A., Energy-efficient secure federated learning for UAV swarms, *7th International Conference on Energy Conservation and Efficiency (ICECE)*, IEEE, 1–5, 2024.
4. Yuan, X., Chen, J., Zhang, N., Fang, X., Liu, D., A federated bidirectional connection broad learning scheme for secure data sharing in Internet of Vehicles, *China Communications*, 18(7), 117–133, 2021.
5. Sasikala, P., Aravinth, T. S., Dharani, K. G., Mohanarathinam, A., Jayachitra, S., IoT-based smart intelligent vehicle systems, *3rd International Conference on Intelligent Sustainable Systems (ICISS)*, IEEE, 1094–1097, 2020.
6. Aruna, S., Padmapriya, T., Saravanan, R., Senthilkumaran, B., Mitigating DDoS attacks: A machine learning approach for enhanced detection and response, *Advances in Nonlinear Variational Inequalities*, 28, 46–59, 2025.
7. Sharma, I., Sharma, A., Gupta, S. K., Asynchronous and synchronous federated learning-based UAVs, *3rd International Symposium on Instrumentation, Control, Artificial Intelligence, and Robotics (ICA-SYMP)*, IEEE, 105–109, 2023.
8. Yazdinejad, A., Parizi, R. M., Dehghantanha, A., Karimipour, H., Federated learning for drone authentication, *Ad Hoc Networks*, 120, 102574, 2021.
9. Prajith, P., Yokesh, V., Artificial intelligence-enabled fully echoed Q-Routing and adaptive directional medium access control protocol for flying ad-hoc networks, *International Journal of Communication Systems*, 38, 1–18, 2025.
10. Valavanis, K. P., Vachtsevanos, G. J., *UAV Fundamentals: Introduction, Handbook of Unmanned Aerial Vehicles*, Springer, 241–242, 2015.
11. Laghari, A. A., Jumani, A. K., Laghari, R. A., Nawaz, H., Unmanned aerial vehicles: A review, *Cognitive Robotics,* 3, 8–22, 2023.
12. Bloise, N., Primatesta, S., Antonini, R., Fici, G. P., Gaspardone, M., Guglieri, G., Rizzo, A., A survey of unmanned aircraft system technologies to enable safe operations in urban areas, *International Conference on Unmanned Aircraft Systems (ICUAS)*, IEEE, 433–442, 2019.
13. Sattler, F., Wiedemann, S., Müller, K. R., Samek, W., Robust and communication-efficient federated learning from non-IID data, *IEEE Transactions on Neural Networks and Learning Systems*, 31(9), 3400–3413, 2019.
14. Wang, Y., Su, Z., Zhang, N., Benslimane, A., Learning in the air: Secure federated learning for UAV-assisted crowdsensing, *IEEE Transactions on Network Science and Engineering*, 8(2), 1055–1069, 2021.
15. Chen, M., Shlezinger, N., Poor, H. V., Eldar, Y., Cui, S., Communication-efficient federated learning, *Proceedings of the National Academy of Sciences of the United States of America*, 118, 1–15, 2021.

16. Zhao, Z., Cumino, P., Esposito, C., Xiao, M., Rosário, D., Braun, T., Cerqueira, E., Sargento, S., Smart unmanned aerial vehicles as base stations placement to improve the mobile network operations, *Computer Communications,* 181, 45–57, 2022.
17. Li, Y., Yuan, D., Sani, A. S., Bao, W., Enhancing federated learning robustness in adversarial environment through clustering Non-IID features, *Computers & Security*, 132, 103319, 2023.
18. Ghosh, A., Hong, J., Yin, D., Ramchandran, K., Robust federated learning in a heterogeneous environment, *arXiv preprint,* arXiv:1906.06629, 1–25, 2019.
19. Yang, H., Zhao, J., Xiong, Z., Lam, K., Sun, S., Xiao, L., Privacy-preserving federated learning for UAV-enabled networks: Learning-based joint scheduling and resource management, *IEEE Journal on Selected Areas in Communications*, 39(10), 3144–3159, 2021.
20. Bonawitz, K., Ivanov, V., Kreuter, B., Marcedone, A., McMahan, H. B., Patel, S., Ramage, D., Segal, A., Seth, K., Practical secure aggregation for privacy-preserving machine learning, *Proceedings of the ACM SIGSAC Conference on Computer and Communications Security*, Association for Computing Machinery, 1175–1191, 2017.
21. Song, Y., Ding, G., Sun, J., Li, J., Xu, Y., Topology tracking of dynamic UAV wireless networks, *Chinese Journal of Aeronautics*, 35(11), 322–335, 2022.
22. Sehito, N., Shouyi, Y., Alshahrani, H. M., Alamgeer, M., Dutta, A. K., Alsubai, S., Nkenyereye, L., & Dhanaraj, R. K., Optimizing User association, power control and beamforming for 6G multi-IRS multi-UAV NOMA communications in smart cities, *IEEE Transactions on Consumer Electronics*, 70, 5702–5710, 2024.
23. Hoang, D. N. M., Truong, V. T., Le, H. D., Long, B. L., Clustered and scalable federated learning framework for UAV swarms, *TechRxiv*, May 4, 2023.
24. Ali Kazmi, S. H., Qamar, F., Hassan, R., Nisar, K., SLFL: A server-less federated learning framework in software defined UAV networks, *IEEE International Conference on Computing (ICOCO)*, IEEE, 339–344, 2023.
25. Fu, M., Shi, Y., Zhou, Y., Federated learning via unmanned aerial vehicle, *IEEE Transactions on Wireless Communications*, 23(4), 2884–2900, 2024.
26. Mohsan, S. A. H., Othman, N. Q. H., Li, Y., Alsharif M. H., Khan M. A., Unmanned aerial vehicles (UAVs): Practical aspects, applications, open challenges, security issues, and future trends, *Intelligent Service Robotics*, 16, 109–137, 2023.
27. Prasanth, A., Sabeena, G., Sowndarya, N., Pushpalatha, N., An artificial intelligence approach for energy-aware intrusion detection and secure routing in internet of things-enabled wireless sensor networks, *Concurrency and Computation: Practice and Experience*, 35, 1–21, 2023.
28. Mothukuri, V., Parizi, R. M., Pouriyeh, S., Huang, Y., Dehghantanha, A., Srivastava, G., A survey on security and privacy of federated learning, *Future Generation Computer Systems,* 115, 619–640, 2021.
29. Yahya, M., Maghsudi, S., Stanczak, S., Federated learning in UAV-enhanced networks: Joint coverage and convergence time optimization, *IEEE Transactions on Wireless Communications*, 23, 1–33, 2023.

8 Advanced Technologies for Federated Learning in Smart Cities and Its Use Cases

Saranya G, Indhumathi G, A. Prasanth, and P. Murugapandiyan

8.1 INTRODUCTION

Rapid growth of urban populations worldwide has catalyzed the development of smart cities using the latest technologies available to enhance the quality life, improve operational efficiency, and ensure sustainable urbanization. Cities generate a voluminous amount of data from sensors, vehicles, and user devices but face challenges in traditional, centralized machine learning (ML) methods in data privacy, security, and scalability. These limitations are compounded with the dynamic nature of heterogeneous data in a smart city.

Federated learning process (FLP), a revolutionary decentralized process of ML, allows local devices (such as sensors and smartphones), which work together to train the models without exchanging any raw data with a central server. In addition to protecting privacy, this paradigm shift lowers latency and bandwidth usage. By facilitating real-time data processing while protecting sensitive data's security and confidentiality, FLP has the potential to revolutionize smart cities. To maximize urban services, a number of progressive cities have started deploying federated learning (FL) systems. For example, initiatives in Singapore and Barcelona concentrate on using FLP to improve healthcare services, lower energy costs, and improve traffic management. These cities serve as examples of how FLP can be smoothly incorporated into municipal infrastructure to handle the escalating issues of real-time decision-making and data privacy.

Federated learning's privacy-preserving methodology is even more important now that regulations are placing more and more attention on data protection. FL offers a scalable, safe basis for the smart cities development in the near future by lowering the possibility of central data breaches and guaranteeing adherence to data privacy laws. The growth of smart cities, where enormous amounts of data are produced by numerous Internet of Things (IoT) devices, sensors, and connected infrastructure, is increasingly dependent on advanced FL technology. Decentralized ML models are made possible by FL, which keeps data at its source while still facilitating

DOI: 10.1201/9781003532323-8

worldwide model training. In smart cities, where sensitive data, like personal information or traffic patterns, must be managed securely, this privacy-preserving strategy is essential.

Block chain, 5G connection, and edge computing are some of the key technologies propelling the integration of FL in smart cities. By ensuring that data processing occurs closer to the source, edge computing lowers latency and BW consumption. For smart city applications, FL can facilitate real-time decision-making when paired with 5G networks, which gives high-speed, low-latency connectivity. Block chain, on the other hand, offers transparent, impenetrable data sharing protocols that guarantee the security and integrity of the FLP.

8.1.1 Background

FL approach enhances the strength of security and data privacy while minimizing communication costs that are attributed to transferring large datasets to central servers. With respect to smart cities, where information is generated from diverse and distributed sources, FL provides a scalable way to deploy ML models across applications.

8.1.2 Motivation

Problems such as increasing data breaches that violate privacy motivate the integration of FL into smart city infrastructures in today's world. In addition to the aforementioned problems, data heterogeneity and the changing nature of urban environments argue for adaptable resilient ML frameworks. This has an edge computing technology, secure multi-party computation (SMPC), homomorphic encryption, block chain, and AutoML, among others, which gives it a great foothold in the development of intelligent, privacy-preserving applications applied in modern society and evolving to meet changing urban smart demands.

8.2 BACKGROUND

8.2.1 Federated Learning Fundamentals

Based on decentralized data processing, FL allows for multiple clients-running smartphones and IoT devices. For example, a global model can be trained without requiring raw data to be communicated by instead exchanging model updates through collaboration. It also solves the problem of keeping private data isolated because it remains on the local device. In normal FLP workflow, the clients download the current modeled global data; then they do local training on their own dataset and return updates for model parameters to a central server for aggregation [1].

8.2.2 Advanced Technologies Enhancing Federated Learning

Several advanced technologies make the functions of FL even more enhanced, thus fit for development in the context of a smart city. Recent developments in blockchain technology have furthered its role in FL through enhanced model update transparency and security. Blockchain creates a decentralized ledger that is immutable while recording transactions and model parameters, which is most important when there is multiparty involvement, such as with the infrastructure of a smart city [27]. Moreover, recent advances on homomorphic encryption – fully homomorphic encryption specifically – mostly allow computations on the encrypted data without necessarily decrypting it first, which also ensures even more secure safeguarding of sensitive data during the lifecycle of model training [28].

Edge Computing: Edge computing can drastically reduce latency associated with data transmission to central servers. Smart cities may have edge devices such as cameras on highways or smart meters on walls, which could possibly compute the local data, train models locally, and make decisions in real time. This will become critical for applications like traffic management where milliseconds make a difference regarding congestion control or accident response. The integration of edge computing with FL enhances the response time and reduces bandwidth consumption in smart cities to a significant extent [2].

Secure Multi-Party Computation (SMPC): It provides the ability of different parties to collaboratively compute a function as inputs without revealing the inputs. In FL, SMPC ensures that sensitive information like health records or financial information never has to be revealed during the entire training of the model. The collaboration between hospitals in smart cities through FL to analyze patient data is an excellent application of SMPC. This also means that common diagnostic models can be designed by the hospitals without exposing data of individual patients so that healthcare becomes smarter and safer at the same time [3].

Homomorphic Encryption: It allows computations to be made directly on the encrypted data without decryption. The methodology ensures that privacy over the data is maintained throughout the ML process [4]. Homomorphic encryption can be used to secure sensitive environmental, traffic, or energy consumption data within smart cities so that data will remain secure even in cases where it is intercepted.

- Block Chain: Block chain technology delivers a tamper-proof decentralized ledger, which is important in multiparty FL collaborations. The FL process becomes much more transparent and secure by recording all model updates and transactions on a block chain. For example, in public safety systems, block chain can ensure that updates of surveillance models are verifiable, preventing tampering, and fostering trust among the stakeholders [5].
- AutoML: It streamlines the processes of model selection and hyperparameter tuning, enhancing both efficiency and performance in FL models [6].

8.2.3 Federated Learning Uses in Smart City Applications

FL provides one with several advantages over traditional machine learning (TML), especially regarding data privacy, scalability, and security. FL is a decentralized model because data doesn't leave the edge devices, hence big improvement in privacy and minimizing the chance of central breach. In addition, the decentralized approach allows for a lower latency due to the processing at the local levels, as opposed to TML, which usually causes a higher latency. But such advantages come with some downsides: mainly increased costs of computation and communication overhead from the need for local training and frequent updates of the model, as shown in Table 8.1.

Several cutting-edge technologies back the incorporation of FL in the smart city infrastructure. Edge computing facilitates the processing of data locally and reduces latency, thus enhancing real-time applications efficiency. Important technologies that advance data security and privacy are SMPC and homomorphic encryption, which, instead of allowing exposure of raw data, perform computation on it. Block chain is a framework that is transparent and secure and can be used for decentralized operation management. The integration of these technologies significantly enhances the capability and security of FL in smart city applications, as illustrated in Table 8.2.

TABLE 8.1
Comparison of Federated Learning Process and Traditional Machine Learning

Feature	Federated Learning Process (FLP)	Traditional Machine Learning (TML)
Data Location	Decentralized (edge devices)	Centralized (data server)
Privacy	High (data remains local)	Low (data aggregated centrally)
Latency	Low (local processing)	High (centralized processing)
Scalability	High (distributed architecture)	Moderate (centralized architecture)
Security	Enhanced (SMPC, encryption)	Vulnerable (central data breaches)
Computation Cost	High (local training required)	Moderate (centralized training)
Communication Overhead	High (model updates required)	Low (data transmission)

TABLE 8.2
Technologies Supporting Federated Learning in Smart Cities

Technology	Description	Application in FL
Edge Computing	Local processing at the data source	Reduces latency, improves efficiency
Secure Multi-Party Computation	Joint computation without sharing raw data	Ensures data confidentiality
Homomorphic Encryption	Encryption allowing computation on cipher text	Preserves data privacy
Block chain	Decentralized, immutable ledger	Provides transparency and security
Automated Machine Learning	Automates model training and optimization	Enhances model performance and efficiency

Together, these technologies make FL systems in smart cities more effective, safe, and scalable while facilitating collaborative innovation across decentralized networks, real-time decision-making, and privacy protection.

FL is used in smart cities and has a variety domain of applications:

- **Traffic Management:** FL will facilitate developing predictive models for traffic flow and congestion without aggregation of sensitive vehicular data [7].
- **Healthcare Services:** FL enables the analysis of shared health data among hospitals with patients' identities being kept confidential [8].
- **Environmental Monitoring:** FL makes possible the central collection of environmental sensor data at different locations to monitor air pollution and other contamination levels [9].
- **Energy Management:** FL helps optimize energy consumption and distribution in smart grids by understanding local usage patterns [10].
- **Public Safety:** FL enhances monitoring systems through real-time anomaly detection without centralizing video data [11].
- **Infrastructure Maintenance**: FL assists the predictive maintenance of urban infrastructure by processing the data from different sensors without sharing the raw data [12].

8.3 FEDERATED LEARNING METHODOLOGIES

8.3.1 Federated Learning with Advanced Technologies

The edge computing has been incorporated with SMPC, homomorphic encryption, blockchain, and AutoML with FL that will improve its usability as well as performance in the city. This fusion helps to bridge the gaps of data privacy, scalability, as well as security efficiency.

8.3.1.1 Edge Computing

It complements FL by providing local computational resources that reduce the latency associated with transmission of data to central servers. This allows FL to have faster model updates and to be instantaneous in its response, especially in the context of traffic management and public safety, which are critical applications [13].

8.3.1.2 Secure Multi-Party Computation (SMPC)

In incorporating SMPC into FL, the respective updates are computed in a secure fashion so that no party can access all the model parameters. This ensures better confidentiality of the training process and this can be rooted to many sensitive applications such as healthcare and military [14].

8.3.1.3 Homomorphic Encryption

It ensures only one to directly execute computations over encrypted data and therefore ensures the privacy of data during model training. This is particularly useful in scenarios where data privacy rules are quite strict [15].

8.3.1.4 Blockchain Technology

Block chain gives an immutable, decentralized ledger for recording updates to the FL model in a way that precludes tampering. This helps to increase the trust of the entities involved in the FL system and its reliability [16].

8.3.1.5 Automated Machine Learning (AutoML)

AutoML in FL will reduce extensive human intervention by automating the processes of hyperparameter tuning and model selection for improving the efficiency of accuracy of the trained models. This makes FL more accessible and scalable [17].

8.3.2 Implementation Framework

The block diagram of the FLP in smart cities is demonstrated in Figure 8.1. It exposes the workflow of how data is handled across various stages, from local data sources to the final global model update. This architecture integrates several advanced technologies to ensure data privacy, security, and efficiency in smart city applications. Each individual component in the block diagram outlines the critical processes and technologies that enable the FL framework to function effectively.

Data Sources

- **Description**: They encompass a range of distributed devices and sensors that have been distributed across different geographical locations, hence giving them their distributed nature in the context of a smart city environment. The sources include traffic cameras, environmental sensors, healthcare devices such as wearables' health trackers, energy meters, public safety, and more.

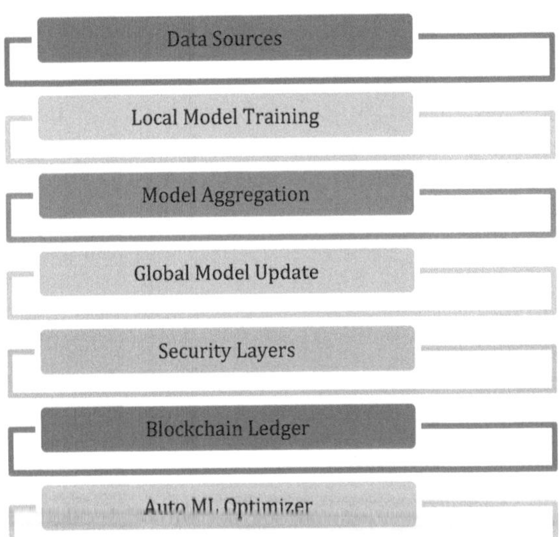

FIGURE 8.1 Federated learning process in smart cities.

Because all of these produce massive data, therefore the distributed location ensures one gets rich information flowing throughout the system.

- **Function**: The primary role of FLP is to have this vast data remain local to its birthplace, and there is no requirement for sensitive data to be transferred to the central server. Instead, raw data that may compromise sensitive information is sent from sources since local models are locally trained at their own sites. This is very important in smart cities where privacy regulations are quite strict, thus minimizing the risk of data leakage. For instance, for healthcare applications, wearable devices can collect patient data and train models on the device without compromising patient privacy. Traffic sensors could monitor traffic patterns while keeping drivers anonymous. Further, a smart city can process the data at the source with reduced costs and security risks of transmission. This can enhance privacy.

Local Model Training

- **Description**: Training of a model by local sources or devices by using their locally available data to train a ML model. In smart cities, it's done in such a way that each device like traffic sensors, air quality monitors, and health devices trains a model and extracts insights from the conditions present in their local conditions.
- **Function**: Locally training models help save bandwidths by sending lesser data on networks. Thus, each traffic sensor can predict congestion patterns on basis of its own immediate surroundings. In the same way, the air quality sensor could train models to predict pollution levels based on local factors like intensity of traffic and industrial activities. Further, in this smart city, by locally training the models.

Model Aggregation

- **Description**: As models are trained for locale on each device or source, the models are collected together and sent to the aggregation server. This has one consequence of reducing the number of transmissions of data, as just model parameters, weights, and bias are being shared while the raw data does not travel. There have also been techniques called federated averaging (FedAvg), which aggregates models and uses them to construct one common global model.
- **Function**: Aggregating all local models to create a global model is, in effect, one of the main procedures in FL. This allows for an all-inclusive model city-wide, using all the distributed devices' insights. Therefore, the aggregation allows for diverse data sources without having any access to the raw data directly. For example, with traffic management, the localized patterns learned by the global model across different intersections could predict more accurately the congested parts of the entire city. In healthcare, models trained on

aggregated local hospital data may be better at diagnosis without sharing any patient-specific information.

Global Model Update

- **Description**: The aggregate local models obtain an updated global model distributed among the participant devices or sources. Thereafter, such participating devices can use the updated global model in the optimization of their local predictions and analysis.
- **Function**: The update to the global model is an essential feedback loop of FL. The insights achieved after aggregating and refining the local into a global model are redistributed back to local devices to benefit each one from the learning that happened in other devices even though they did not access the data of the other devices. For example, a traffic sensor located in one part of the city may use information related to congestion in some other part of the city in order to improve the precision of its local forecasts against a larger citywide picture. Likewise, hospitals can exploit a global healthcare model to have an enhanced ability to correctly diagnose, depending on aggregate learning from a pool of data. That means cyclic update of models in the course of smart city systems ensures continuous improvement and adaptation.

Security Layers

- **Description**: This FLP involves multiple security layers that need to be implemented in order to support the protection of data integrity as well as confidentiality. SMPC, homomorphic encryption, and differential privacy are some of the technological foundations for protecting data in the training as well as aggregation phases.
- **Function**: FL is mainly critical in smart cities as sensitive information involving personal health data or public safety footage is deployed. SMPC allows multiparty computation in a secured way, which means multiple parties can jointly compute model updates without disclosing their individual data. In this way, no party would be able to gain access to raw data from the other party even at the time of aggregation. Homomorphic encryption supports running computations on encrypted data, hence the computations of sensitive information are done without decryption at any point. Differential privacy further enhances this protection, introducing noise in the updates such that it is impossible to retrieve even a single piece of data and trace it back to the individual. Together, these security measures ensure that the private data of smart city residents is protected while ensuring transparency and trustworthiness in the FLP. Security layers prevent unauthorized data access and model updates, thus ensuring sensitive data remains private and protected during FL.

Block Chain Ledger

- **Description**: Blockchain plays a critical role in FL by ensuring a tamper-proof record for all the model updates and transactions. Decentralized ledger ensures the process of FL – every step from local model training to global model updates is transparent and traceable.
- **Function**: Blockchain introduces the immutable and decentralized ledger system that provides transparency and trust to the process of FLP. In the smart city where government agencies, companies, and public institutions might be involved in participating in FL, there must be no chance that any party can manipulate the FLP and model updates. Accordingly, the blockchain feature ensures accountability through permanent records of all transactions that can be verified by every individual. It is particularly helpful for most applications in the context of public safety where accuracy of video surveillance data as well as updates in models needs to be trusted to the core. With the assurance of blockchain, smart cities can ensure collaboration and trust among stakeholders about the FLP.

AutoML Optimizer

- **Description**: AutoML tools are used for automating the iterative processes necessary to train ML models. These include model selection, hyper parameter tuning, and feature selection that are the primary requirements for optimizing the performance of FL models.
- **Function**: AutoML has played a very important role in the federating learning task process that would be quite time-consuming and inefficient when tuned by a human being, more so at large scales of application in smart cities. It automatically determines the best model by optimizing hyperparameters with guarantee of fit models suitable to an application profile of different applications within the smart city. Best is the traffic model that predicts traffic. AutoML can tune, without human intervention, learning rates and batch sizes to maximize traffic prediction model performance. Further instances are the optimization of diagnostic models used in healthcare applications, with the choice of best algorithms and features to predict diseases. The automation enhances the overall accuracy and efficiency of the models and makes FL more accessible to the smart city administrators who do not possess advanced expertise in ML.

8.3.3 Applications of Use Case Development

8.3.3.1 Traffic Management and Its Maintenance

Traffic congestion is a major and important concern in many cities, causing significant economic damage and environmental consequences. Combined learning can be used to develop predictive traffic models that analyze data from traffic cameras and distributed sensors. These models can predict congestion and adjust the timing of traffic lights or suggest optimal delay strategies. For example, smart cities such as

Hangzhou in China have successfully implemented AI-powered traffic systems, and storing FL vehicle data locally and privately could enhance these systems greatly and still improve overall traffic forecasting [18].

8.3.3.2 Healthcare Services

Healthcare is a sensitive area where statistics privacy is essential. Hospitals and healthcare carriers in smart towns can collaborate using FL to create powerful diagnostic models without compromising patient privacy. Each sanatorium can teach a neighborhood model the usage of its patient records and send the simplest model updates to an international version, which aggregates the findings from different assets. This technique allows accurate diagnostics and treatment plans across hospitals, while adhering to strict privacy policies like health insurance portability and accountability act (HIPAA) and general data protection regulation (GDPR). Hospitals and healthcare companies collaborate to teach a shared diagnostic version of the usage of affected person information saved regionally. The federated technique guarantees patient privacy at the same time as permitting the improvement of correct diagnostic gear [19].

8.3.3.3 Environmental Monitoring

Cities with big populations face challenges associated with pollution and environmental degradation. FL lets clever towns to screen air first-rate, temperature, and pollution tiers in real-time using information from a network of sensors. These sensors can educate models locally to discover pollution traits and provide actionable insights to metropolis planners while not having to percentage touchy environmental data, thus keeping privacy and safety. Distributed environmental sensors acquire statistics on air quality, temperature, and humidity. FL fashions analyze these statistics regionally to discover pollutants traits and offer actionable insights to town planners [20].

8.3.3.4 Energy Management

Smart meters and energy garage devices use FL to optimize power distribution and intake styles. The federated models help in balancing the strength load efficiently across the town [21]. In the context of electricity control, gaining knowledge can substantially enhance the performance of smart grids by reading strength consumption patterns at the nearby degree without sharing touchy usage data. Smart meters deployed in city families and industries can collaboratively teach predictive models to optimize electricity distribution and come across anomalies, consisting of power theft or gadget malfunction. These models help in preserving grid stability and improving power efficiency, while concurrently addressing privacy worries [22].

8.3.3.5 Public Safety

Public safety in smart cities benefits immensely from FL via allowing real-time video surveillance and anomaly detection. Cameras and sensors dispersed across the city can locally procedure visible records to discover capability threats, together with unauthorized intrusions or injuries, without sending uncooked photos to a valuable server. This decentralized technique ensures that privateness is preserved while leveraging collaborative learning fashions to decorate security [23].

8.3.3.6 Infrastructure Management

Smart city infrastructure, together with roads, bridges, and water structures, calls for nonstop tracking and protection. FL may be employed to increase predictive renovation fashions via analyzing statistics from dispensed sensors. For example, sensors positioned on bridges or roads can train models locally to expect structural failures, taking into consideration timely upkeep at the same time as maintaining the statistics stable.

8.4 ADVANTAGES, CHALLENGES, AND FUTURE SCOPE OF FL

8.4.1 Advantages of FL in Smart City Environments

FL has various advantages for clever towns, together with expanded statistics privacy, decreased communique price, and more scalability. By maintaining information localized, FL reduces the risk of statistics breaches and assures adherence to information safety requirements. Additionally, FL's decentralized nature permits for greater scalable and resilient system mastering fashions, able to adapt to the dynamic surroundings of smart towns [24].

8.4.2 Challenges and Solutions

Despite of its advantages, in the context of smart cities, FL confronts the subsequent difficulties:

- **Data Heterogeneity**: Challenges in federated studying are the heterogeneity of fact sources. Different gadgets, sensors, and systems in a clever city may also generate information in varying codecs and systems, which could result in inconsistencies inside the trained models. Solutions to this task involve the usage of AutoML techniques to standardize records codecs and optimize fashions throughout diverse sources. Additionally, version aggregation techniques, like FedAvg, can help mitigate the effect of data variability by ensuring that neighborhood updates are uniformly mixed into the global model [25].
- **Communication Efficiency**: FL requires frequent communique among neighborhood devices and the significant server to replace global models that could stress network sources. To cope with this, aspect computing can be leveraged to perform greater computations locally, decreasing the frequency of communication with the relevant server. Additionally, the compression method is used to reduce the dimensions of version updates, as a consequence reducing bandwidth requirements and ensuring smoother actual-time operations in smart city programs [26].
- **Security Threats**: Federated mastering, regardless of its decentralized nature, is not proof against security dangers. Malicious actors may additionally try

to compromise the device by way of injecting adversarial facts or appearing version poisoning assaults that could degrade the performance of the global version. Solutions like homomorphic encryption and differential privateness provide protection via making sure that neighborhood records and version updates are encrypted, even in the course of computations. Moreover, strong aggregation techniques, along with Byzantine fault tolerance, can hit upon and discard malicious model updates, safeguarding the general integrity of the FL system [27].

- **Adversarial Attacks and Data Poisoning**: One of the crucial security challenges in FL is opposed attacks, where malicious actors inject poisoned records or version updates to degrade the model's performance. Another form of assault, statistics poisoning, entails altering the education facts to control the worldwide model's getting to know procedure. Solutions to those threats include the use of strategies like homomorphic encryption, which allows computations on encrypted records, and differential privateness, making sure that personal information factors continue to be confidential at some point of version updates. Additionally, strong aggregation techniques consisting of Byzantine fault tolerance can help discover and neutralize malicious version updates, improving the resilience of FL in smart metropolis packages [28, 29].

8.4.3 Future Directions

Future studies have to focus on improving the interoperability of superior technologies with FL, developing robust safety protocols, and exploring novel programs within smart towns. Additionally, FL can be similarly stronger in clever city areas with the incorporation of upcoming technologies like 5G and IoT [30]. Introduction of 5G networks is poised to further boost up the adoption of federated gaining knowledge in smart cities. With extremely low latency and excessive bandwidth, 5G networks will allow faster statistics transmission and version updates, making FL structures greater responsive. Coupled with the proliferation of IoT devices, which continuously generate tremendous quantities of records, 5G will permit actual-time FL programs in regions such as autonomous vehicles, smart energy grids, and public health structures. The combination of these technologies will lead to more resilient and adaptable clever city ecosystems [31].

The destiny of FL in smart cities lies in the integration process of emerging technologies along with 5G and IoT. With 5G, federated getting to know can achieve even lower latencies and better bandwidth, enabling actual-time packages, which include independent motors, smart energy grids, and public health structures, to flourish. Ultra-low latency of 5G networks will permit FLP systems to replace models instantly, improving the responsiveness of smart city infrastructures.

5G Networks and FL: 5G networks, with their high data transfer rates and occasional latency, will beautify the scalability of federated studying applications in smart towns. The excessive bandwidth supplied via 5G will support the continuous drift of data from hundreds of thousands of IoT gadgets, permitting FL to swiftly update global models. Autonomous motors, for instance, will take advantage of 5G-enabled

FL with the aid of receiving real-time visitor updates and predictive fashions that help avoid injuries or congestion.

IoT Integration: The IoT is another critical area, a good way to pressure the boom of federated gaining knowledge in clever cities [32]. As greater devices come to be interconnected, IoT sensors will generate vast amounts of records that may be processed domestically through the use of FL, reducing the need for centralized information storage and processing. The mixture of IoT and FL will permit clever cities to optimize operations in actual time, whether or not it is energy control, public safety, or environmental monitoring.

AI and FL: The integration of AI with federated studying will free up new possibilities for clever cities. AI-driven systems can automate numerous city operations, such as site visitors management, waste management, and catastrophe reaction, by means of schooling system studying fashions collaboratively across a couple of devices. This method guarantees that AI structures operate effectively without compromising privateness or protection.

8.5 CONCLUSION

FLP is a noteworthy improvement in the subject of decentralized gadget learning, supplying a private and steady technique suitable for the elaborate and ever-converting smart metropolis surroundings. By integrating advanced technologies, including facet computing, SMPC, homomorphic encryption, blockchain, and AutoML, FL can successfully address the difficulties with privacy, scalability, and protection of records. The realistic applications of FL in visitor management, healthcare, environmental monitoring, strength management, public safety, and infrastructure protection show its transformative capacity in optimizing urban operations and enhancing first-rate of life. Adoption of FL can be crucial to constructing resilient and sustainable urban ecosystems as smart towns expand.

REFERENCES

[1] Konecny, J., McMahan, H.B., Yu, F.X., Richtárik, P., Suresh, A.T., and Bacon, D. Federated learning: Strategies for improving communication efficiency, *arXiv preprint*, 1–10, 2016. https://doi.org/10.48550/arXiv.1610.05492.

[2] Shi, W., Cao, J., Zhang, Q., Li, Y., and Xu, L. Edge computing: Vision and challenges, *IEEE Internet of Things Journal*, 3, 637–646, 2016.

[3] Bonawitz, K., Ivanov, V., Kreuter, B., Marcedone, A., McMahan, H.B., Patel, S., Ramage, D., and Seth, K. Practical secure aggregation for privacy-preserving machine learning, *Proceedings of the 2017 ACM SIGSAC Conference on Computer and Communications Security*, 1175–1191, 2017.

[4] Aruna, S., Padmapriya, T., Saravanan, R., and Senthilkumaran, B., Mitigating DDoS attacks: A machine learning approach for enhanced detection and response, *Advances in Nonlinear Variational Inequalities*, 28, 46–59, 2025.

[5] Zheng, Z., Xie, S., Dai, H., Chen, X., and Wang, H. Blockchain challenges and opportunities: A survey, *International Journal of Web and Grid Services*, 14, 352–375, 2018.

[6] Hutter, F., Kotthoff, L., and Vanschoren, J. *Automated Machine Learning: Methods, Systems, Challenges*, Springer, 2019.
[7] Yang, Q., Liu, Y., Chen, T., and Tong, Y. Federated machine learning: Concept and applications, *ACM Transactions on Intelligent Systems and Technology*, 10, 1–19, 2019.
[8] Rieke, N., Hancox, J., Li, W., Milletari, F., Roth, H.R., Albarqouni, S., et al. The future of digital health with federated learning, *NPJ Digital Medicine*, 3, 119, 2020.
[9] Wang, S., Tuor, T., Salonidis, T., Leung, K.K., Makaya, C., He, Y., and Niyato, D. Blockchain-enabled federated learning for vehicular networks, *IEEE Transactions on Intelligent Transportation Systems*, 22, 1523–1534, 2021.
[10] Yang, Q., Wang, S., and Zhu, X. Energy-efficient federated learning for smart grids, *IEEE Transactions on Smart Grid*, 12, 2261–2272, 2021.
[11] Zhang, Y., Qiu, M., Tsai, C.W., Hassan, M.M., and Alamri, A. Health-CPS: Healthcare cyber-physical systems assisted by cloud, fog, and edge computing, *IEEE Systems Journal*, 13, 637–652, 2019.
[12] Lin, X., Li, C., Zeng, L., and He, H. Federated learning-based predictive maintenance for smart infrastructure, *IEEE Internet of Things Journal*, 8, 12345–12358, 2021.
[13] Bonomi, F., Milito, R., Zhu, J., and Addepalli, S. Fog computing and its role in the internet of things, *Proceedings of the First Edition of the MCC Workshop on Mobile Cloud Computing*, 13–16, 2012.
[14] Mohassel, P., and Zhang, Y. SecureML: A system for scalable privacy-preserving machine learning, *Proceedings of the 21th ACM SIGSAC Conference on Computer and Communications Security*, 19–38, 2014.
[15] Halevi, S., and Shoup, V. Implementing privacy-preserving machine learning with homomorphic encryption, *IEEE Transactions on Knowledge and Data Engineering*, 32, 1234–1245, 2020.
[16] Liu, Y., Wang, S., and Niyato, D. Blockchain-based secure federated learning in mobile edge networks, *IEEE Internet of Things Journal*, 7, 8905–8914, 2020.
[17] Feurer, M., and Hutter, F. Hyperparameter optimization, *Automated Machine Learning*, 3, 3–33, 2019.
[18] Ma, X., Liu, Q., Li, B., and Xu, D. Federated learning for intelligent traffic management: A survey, *IEEE Transactions on Intelligent Transportation Systems*, 23, 5678–5690, 2022.
[19] Zhao, Y., Li, M., Lai, L., and Li, H. Privacy-preserving federated learning for healthcare, *IEEE Journal of Biomedical and Health Informatics*, 25, 1234–1245, 2021.
[20] Chen, T., Zhang, Y., and He, H. Federated learning for environmental monitoring in smart cities, *Environmental Science & Technology*, 54, 6789–6798, 2020.
[21] Kumar, N., Singh, A., and Gupta, R. Federated learning for energy management in smart grids, *IEEE Transactions on Smart Grid*, 12, 4567–4578, 2021.
[22] Yang, Q., Liu, Y., Chen, T., and Tong, Y. Federated machine learning: Concept and applications, *ACM Transactions on Intelligent Systems and Technology*, 10(2), 1–19, 2019.
[23] Wang, S., Tuor, T., Salonidis, T., Leung, K.K., Makaya, C., He, Y., and Niyato, D. Blockchain-enabled federated learning for vehicular networks, *IEEE Transactions on Intelligent Transportation Systems*, 22(3), 1523–1534, 2021.
[24] Shokri, R., and Shmatikov, V. Privacy-preserving deep learning, *Proceedings of the 22nd ACM SIGSAC Conference on Computer and Communications Security*, 1310–1321, 2015.
[25] Li, T., Sahu, A.K., Talwalkar, A., and Smith, V. Federated learning: Challenges, methods, and future directions, *IEEE Signal Processing Magazine*, 37, 50–60, 2020.

[26] McMahan, H.B., Moore, E., Ramage, D., and Hampson, S. Communication-efficient learning of deep networks from decentralized data, *Proceedings of the 20th International Conference on Artificial Intelligence and Statistics*, 1273–1282, 2017.

[27] Bagdasaryan, E., Veit, A., Hua, Y., Estrin, D., and Shmatikov, V. How to backdoor federated learning, *Proceedings of the 6th International Conference on Learning Representations*, 2938–2948, 2019.

[28] Fung, C., Yoon, C. J. M., and Beschastnikh, I. The limitations of federated learning in sybil settings. *Proceedings of the 23rd International Symposium on Research in Attacks, Intrusions and Defenses (RAID 2020)*, 301–316, 2020.

[29] Zheng, Z., Xie, S., Dai, H., Chen, X., and Wang, H. Blockchain challenges and opportunities: A survey. *International Journal of Web and Grid Services*, 14(4), 352–375, 2018.

[30] Gentry, C. Fully homomorphic encryption using ideal lattices. *Proceedings of the 41st Annual ACM Symposium on Theory of Computing (STOC 2009)*, 169–178, 2009.

[31] Andrews, J. G., Ghosh, A., and Ratasuk, R. What will 5G be? *IEEE Journal on Selected Areas in Communications*, 32(6), 1065–1082, 2014.

[32] Aruchamy, P., Lavina B., and Devi, S. An energy-aware link fault detection and recovery scheme for QoS enhancement in Internet of Things-enabled wireless sensor network, *Computers and Electrical Engineering*, 123, 1–17, 2025.

9 Federated Deep Learning for Cyber-Physical Systems in Real-World Scenarios

Princy K T M, P. Manjula, K. Saranya, and G. Uthradevi

9.1 INTRODUCTION

Cyber-physical systems (CPS) are one of the intelligent systems that amalgamate physical and logical devices. Physical systems can interact with the environment, and the logical system is responsible for processing and making intelligent decisions based on the physical system's interaction with the environment. In modern days CPS are an important component of smart infrastructure and autonomous systems. CPS are used in almost all fields, including autonomous physical systems, manufacturing, medical, military, process control, power generation, agriculture, energy distribution, energy conservation, schools, colleges, smart cities, hospitals, railways, research centers, and everywhere [1].

CPS have the immense capacity to interact with the environment and process data to make intelligent systems. CPS connect with the Internet of Things (IoT), industrial IoT, Internet of everything, machine to machine, and trillion sensors [2–4]. CPS help in many ways. For example, a farmer can pour the pesticides with the help of an intelligent system without directly interacting with the pests and pesticides. With the integration of IoT, artificial intelligence (AI), and CPS, the traditional healthcare system became a smart system [5]. The introduction of IoT helped to maintain drastic changes in the medical system.

CPS integrate various IoT devices along with other components such as computers and cloud services. After the introduction of and deep learning (DL) and machine learning (ML), the system can make quick decisions based on the data. These decisions seem like expert human decisions based on their professional experience. Error in these decisions is very less. This technology helps the medical cyber system to do a lot of different intelligent tasks [6]. However, the privacy of the data is the most significant factor for achieving intelligent systems. All these technologies focused on efficiency and avoided security and privacy infringements.

DOI: 10.1201/9781003532323-9

In DL and ML, the system learns from different functionalities and features from the dataset [7]. ML and DL typically help to improve the system. Nevertheless, this learning method uses data that adds vulnerability to the model. The current chapter's organization is as follows. Section 9.2 introduces CPS and Section 9.3 deals with medical cyber-physical systems (MCPS). Section 9.4 provides the architecture of federated learning (FL)-based MCPS. Section 9.5 deals with model training and algorithms. Section 9.6 demonstrates the challenges faced by FL.

9.2 CYBER PHYSICAL SYSTEM

The CPS architecture typically contains software and hardware components. Figure 9.1 depicts the architecture of the CPS. Physical processing devices, sensors, communication networks, actuators, and control algorithms are general components of the CPS architecture [8]. Physical systems are the components that interact with the environment. Sensors are connected to physical systems to collect data. The network components act as the bridge to transfer the data between the physical and cyber domains.

The computational node of the CPS processes data from the different nodes. Actuators control the physical process based on computational needs. The control algorithm analyses the data and makes decisions. All learning and decision-making happen inside the cyber domains with the help of computational algorithms. CPS can possess different features like networking, data analysis, computation, and heterogeneous data processing. All these features of CPS make the system to perform smarter and intelligent computations [9].

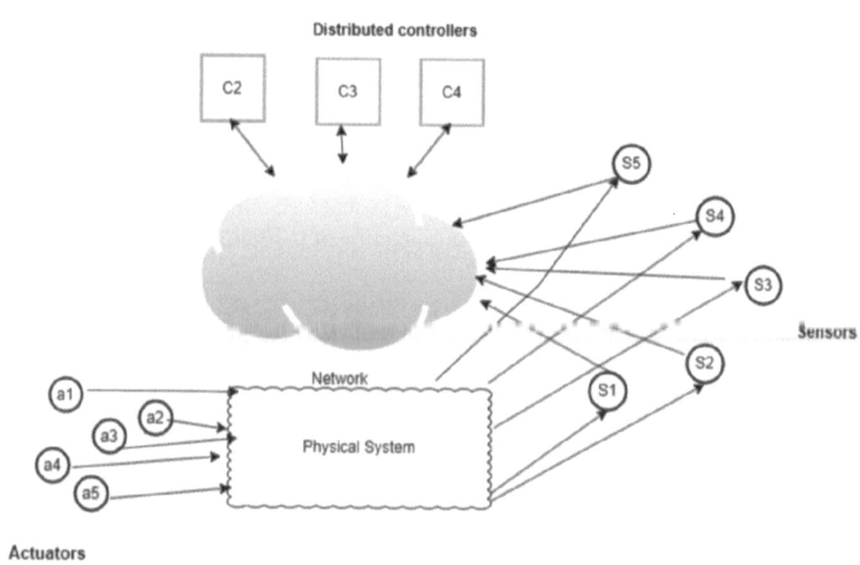

FIGURE 9.1 Basic structure of the CPS technique.

9.3 MEDICAL CYBER-PHYSICAL SYSTEM

MCPS are cyber systems that integrate medical devices. They include medical components like X-rays, ECG, MRI, and doctor's prescription generation devices. They are progressively used in hospitals, and they generate lots of data. These technologies are essential to contemporary medical procedures because they make it easier to handle and analyze the data. CPS data flow is described in Figure 9.2. These data flows are also applicable to the MCPS [10].

The architecture of the MCPS is exposed in Figure 9.3. Various medical settings and healthcare sensors make up the MCPS's physical components. The cyberspace of the MCPS is the brain of the MCPS; it stores, processes, and manages the user's security and health information [11]. In the medical field, health records include national health information of patients, patient personal health information, and data related to operations room procedures, etc. The advantages of the MCPS include the integration of different networks with different features. Human system interactions help to take data directly and add human perceptions into the MCPS systems [12]. Scalability and flexibility of adding new features and devices into the system improve adaptability to become and work as a large-scale information system.

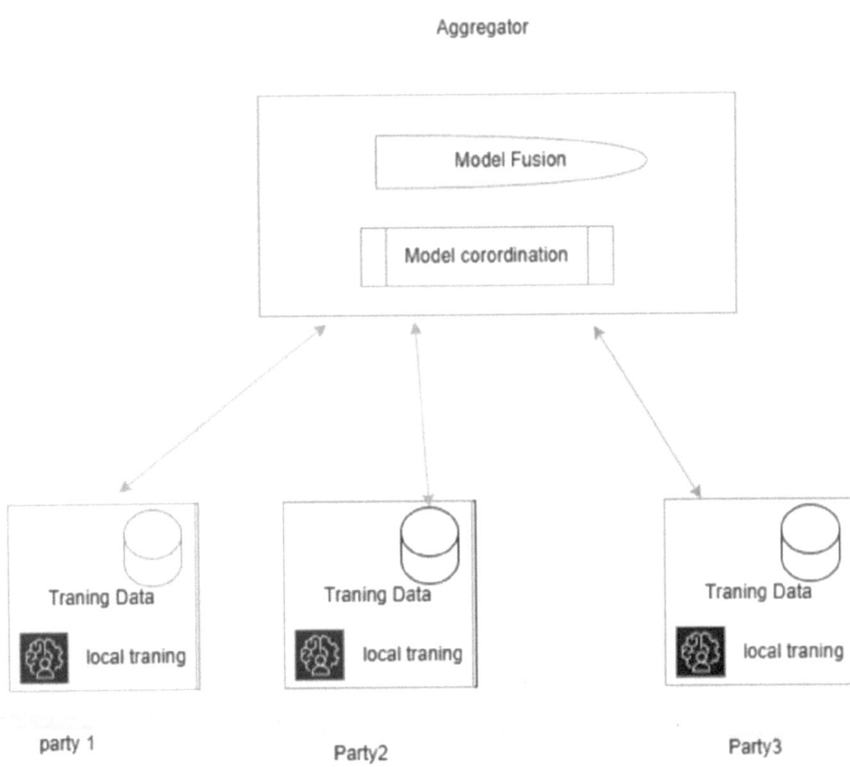

FIGURE 9.2 General architectures of federated learning.

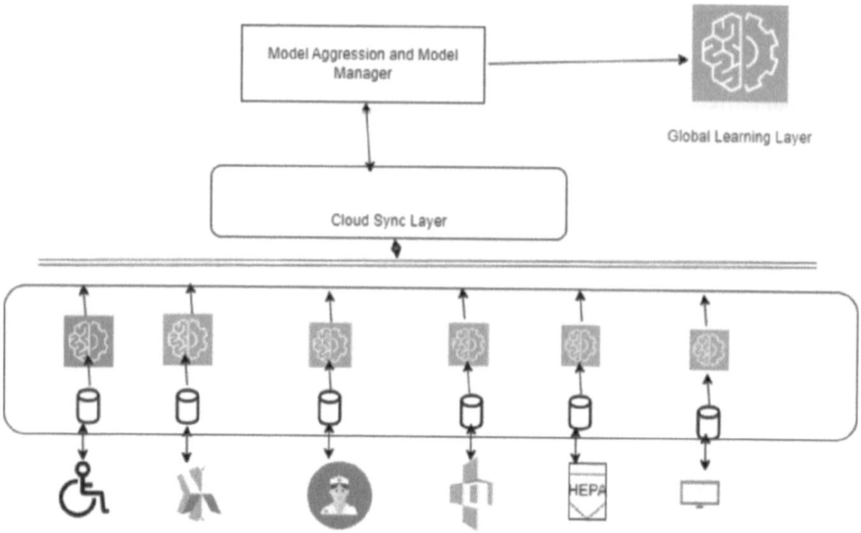

FIGURE 9.3 Architecture of federated learning in MCPS.

Even though it provides significant flexibility and advantages to the healthcare system, data privacy and security remain important concerns. Data privacy and security issues are important concerns. Whenever the MCPS system is active it produces large amounts of data and processes different electronic records. All these data are the sources of the new inventions, and all this data should be secured. It contains lots of sensitive information like patient personal and disease details, diagnosing details, and also the hospital's new findings and new procedure methods. For example, medical sensitive data contains medical history, lab results, prescribed medicine, lifestyle details, etc.

These days all these medical records are stored in electronic databases, and it is connected to the MPS. These electronic records always face security challenges. If these data are out, it creates social problems and affects the security of the medical system and hospitals. It needs to maintain more security and confidentiality. But with the introduction of data science and AI, all these data security and privacy were lost.

9.4 FEDERATED LEARNING

FL training includes four main steps: they are the initialization of the model, selecting or identifying the parties or clients in the learning process, training the model in local devices, aggregating all the models and collecting, and integrating to the main devices in the centralized server [13]. Each of these steps includes tedious tasks in this model. Aggregation is the difficult one because of the heterogeneity of the data. FL follows the logic that there are a number of three users (part 1, part 2, part 3, and part n as shown in Figure 9.2). The parties are involved in the learning process, and everyone carries theory data and none of them can access others' data.

FedAVg is a well-known FL algorithm. FedAvg is a well-known FL algorithm. Since FL handles different types of heterogeneous data, a more generalized approach called FedGAvg has been developed to address this challenge. [14]. In the local Global Federated Averaging Method, local devices learn and extract higher-level representations from raw data prior to the operation of a global model on the representations from other devices. The primary issue with FL is data heterogeneity. These data are distributed identically and independently.

Data and system heterogeneity issues lead to a lot of incorrect results and disadvantages. Poisoning attacks are the other types of challenges that the FL system would face. Data poisoning and model poisoning attacks are the two main types of poisoning attacks. An intruder handles the data utilized by a user that causes a false advertisement to the global system. In the case of model poisoning attack, the intruder sends the wrong updates to the global system. These types of attacks affect the robustness and accuracy of the FL model. By establishing anomaly detection and robust aggregation, FL systems can protect [15].

9.5 IMPACT OF FL IN MCPS

Normal ML model learning happens in the centralized server or system. The decentralized approaches of ML or DL are known as FL. FL was introduced in 2017 by Google. FL enables different organizations to work together without sharing their trade information and secrets. In many cases, poor network connection and data from different sources connection are difficult. FL overcomes this difficulty while working in the local node itself [16].

FL architecture can be done in many ways. The general form of FL is represented in Figure 9.2. In this, the aggregator component is called "the coordinator". It controls the systems and coordinates. Here, the parties are the different clients or different organizations or different devices that are participating in the learning process while contributing the data. Aggregators collect the model updated from the different models and then merge it to form a fusion model called "a global model". The result of this system is again distributed to the next step for the next stage of learning to update the parameters or weights or different outputs. This process will repeat until all the learning converges. The result of the FL model depends on all the party's data without collecting data from the parties [17].

Different types of architectures developed for the implementation of FL in CPS. Components of CPS are the intersection of physical systems and virtual systems. It includes mainly three components: a learning participant, an aggregator, and a communication network. Aggregator functions are to aggregate and manage the FL components. The FL does not follow the distributed processing architecture. Learning happens in the local parties where the aggregator collects all the information from the local parties and updates the models. Network and communication act as a chance to interact with the local parities of the aggregator. The FL data distribution is not handled by the aggregator or the central. The central system is not aware of how much data is stored in the parties and their dependencies. Some parties share more data compared to the other parties.

9.5.1 Architecture of FL-Based MCPS

The detailed structure of FL-based MCPS is demonstrated in Figure 9.3. It is the intersection of virtual and physical components. It is difficult to separate the physical and virtual components separately in MCPS. Equations 9.1 and 9.2 represent the mathematical formulation of the FL in MCPS. FL in MCPS happens with the help of different physical and virtual components. Sensors, medical devices, actuators, and other physical components are all included in the medical CPS physical component. Virtual entities include different software and learning algorithms, computational algorithms, and cyberspace [18].

$$\text{Cyber\_Physicalsystem} = \text{Phy}_{\text{com}} \cap \text{Virtual}_{\text{Comp}} \tag{9.1}$$

$$\text{Phy\_Comp} = \sum n = 1\,\text{Mpi} \tag{9.2}$$

$$\text{Virtual\_Comp} = \sum n = 1\ \text{Mpi} \tag{9.3}$$

Virtual components of the MCPS are different computational algorithms, communication technologies, patient monitoring software, etc. Computational algorithms are like decision trees, Bayes models, and classification models. Cloud computing technologies include the technologies for storing and processing data on the cloud. Patient monitoring systems include different software to store and process the patient's information centrally. Physical components of the FL in MCPS include different hospital and clinical physical and visible contents like computers, patients, doctors, and nurses. It also has different edge devices like IoT mobile devices and embedded systems [19].

9.5.2 Factors Considering for FL Training

FL follows the principle that different parties can participate in the local training model, but each one is not aware of that model. The learning process involves n number of clients with n number of datasets [20]. One of the major features of FL training is that it can train with heterogeneity in the data such as medical images, clinical data, and different types of data. FL is capable of training data structures that are independently and identically distributed. The main issue associated with identically distributed data is skewness in the data. It is the variation of labels in the same dataset with different names.

Feature distribution skew occurs when some client datasets include a broader or more diverse set of features compared to others. Using data augmentation and updating the parameters consistently can help to overcome these issues [21]. Normalization techniques on the dataset can also deal with reducing the skewness in the distribution of the dataset. Quantity skewness is the other issue faced while training the model. Quantity skewness means quantity changes in the dataset. Clients can have the tendency to train using unequal amounts of data because of several reasons. This amount of variations in the dataset significantly contributes to the final evaluation of the model [22].

9.5.3 Model Aggregation Phase

Model aggregation is the process of gathering model parameters from different parties where every cycle of learning model parameters is collected. By using all these parameters, the final global updation is accomplished. Network fracture is an important aspect to consider while training the dataset. The central server is the aggregator. Parameter-based model aggregation and output-based aggregation are the types of aggregation. In parameter-based aggregation, local model trainable parameters, such as weights and gradients of deep neural networks, are joined together [23].

The trainable parameters of local systems (weights and gradients) are combined in parameter-based aggregation. Local systems are iteratively trained on their concern datasets throughout each communication round, and the new gradients or parameters are transferred for aggregation [24]. The Fedmask approach is the most commonly used in output-based aggregation methods. Fedmask can address the challenges of the heterogeneity of the data. Only a portion of the trained model is shared with the global model in the Fedmask model [25].

Scaffold is the other aggregation algorithm. This algorithm can control the technology heterogeneity of the local devices. It adds two additional variables for client control to the algorithm such as control variable Client C and control variable Client K. Typically, there are numerous algorithms employed for aggregation. FedAvg is the oldest algorithm for aggregations. In FedAvg, groups of clients are randomly selected in the round for aggregation. At the stage of aggregation, the parameters of each client are taken, and the average of the parameters is taken into consideration in the model [26].

FedAvg has the drawback of dealing with the optimized parameters taken from the local model. To mitigate this risk, the FedPRox approach is developed. It reduces the localized optimization problems. The other algorithm technique is MOON that will diminish the difference between the local and global systems. This can be accomplished by incorporating a model contrastive loss as a regularization term.

9.5.4 FL-Based MCPS Algorithm

The following steps expose the sequential steps of the FL-based MCPS algorithm. These steps provide an overview about the model training and gradient calculation.

1. Initialize the Global Model
   - Make the first global model parameter G_0 as the starting value. Let D represents the local data on each device.
2. Ensure the Gradient of the Local Loss
   - Evaluate the local loss function. It should be optimized with respect to local data.

3. Initialize the Local Model
    - Set the local model *Lm* to the starting value of the global model.
4. Local Model Training
    - To train this system, the local data *D* is utilized.

$$L_m \leftarrow \text{TrainLocalModel}\left(L_m, D\right)$$

5. Compute Local Gradient
    - After training, compute the local gradients based on the updated local model:

$$\text{local\_gradient} \leftarrow \text{ComputeGradients}\left(L_m, D\right)$$

6. Update the Server
    - Return the updated model parameters or calculated gradients to the server. It can be aggregated across all involved devices.

9.6 TYPES AND CHALLENGES OF FL-BASED MCPS

In recent decades, different types of FL methods have been established. Data partitioning, privacy mechanisms, ML models, and communication architecture can all be used to categorize FL in MCPS [27]. Based on the data partition, federated transfer learning, vertical FL, and horizontal FL are formulated, which is illustrated in Figure 9.4. In horizontal FL, the user's features of the dataset overlap a lot, but the users are different. For example, the medical field hospitals and research hospitals have a similar dataset of a disease. Still, the patients are different and their other disease features or lifestyle conditions vary. However, the number of organizations involved in this model training increases the quality of the model by adding more features to the dataset. In vertical FL, the user's features normally do not overlap much but the users are overlapping.

In simple terms, data of the same user's division can be separated into columns or features in the training process. In the medical field, different hospitals collect

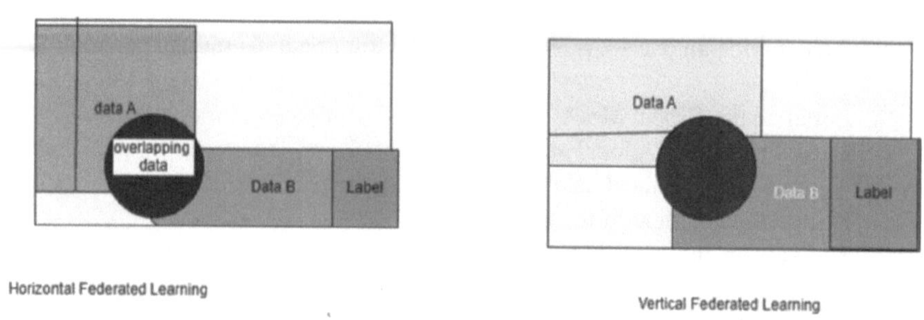

FIGURE 9.4 Types of federated learning.

and train the information of specific patients or groups of patients where training is manipulated by using the features in different devices. Classification and linear regression algorithms are particularly used for vertical FL algorithms [28]. Federated transfer learning methods are methods that use already existing models. This model is not used directly due to its existing structure. All these models have already been trained by using the existing large dataset. Thus, while using these models, the parameters are only updated.

Based on the privacy issues, three types of FL are exploited [29]. They are model aggregation FL, holomorphic encryption FL, and differential privacy FL. Model aggregation is an important step in FL. Model aggregation summarizes all different parameters from all parties to avoid transmitting the original data in the training process. Besides, the cross-device and cross-silo are other types of FL. Cross-devices are presumed based on the network size whereas the cross client is accomplished based on the party or client size. The cross-silo FL is considered when FL is utilized by different organizations with bigger sizes.

9.6.1 Vulnerabilities of FL-Based MCPS

In this section, the difficulties and vulnerabilities of FL as well as their way to reduce these challenges are listed. Communication cost, system heterogeneity, and statistical heterogeneity are the main obstacles to FL-based MCPS [30].

9.6.1.1 Security Threats in Federated Learning

In the proper FL setup, only the aggregator and parties can have access to data either partially or fully. There are two types of threats generally FL model faces: inside threats and outside threats. Inside adversaries manipulate the model while training and outside adversaries try to manipulate the data. The inference threats are the threats in which the outside tries to extract private information from the data. These types of threats cause misclassification and wrong prediction, which decreases the efficiency of the model.

9.6.1.2 Byzantine Attacks

In Byzantine attacks, the attacker controls multiple users by conveying the updated parameters to the global model. The attackers continuously generate conflicting messages that create confusion and affect the various parts of the system. Further, it sends erroneous updates to the global model.

9.6.1.3 Attack on Training Process

During training the model, the attacker attempts to learn, influence, and compromise, which affects the training process performance. Manipulation attack generally serves as the external parties or the aggregator. One of the biggest challenges with manipulation attacks is that manipulation attacks are difficult to detect and categorize.

9.7 CONCLUSION

FL has generated a lot of curiosity in creating scalable and privacy-enhanced healthcare apps and services. An overview of FL in the medical system was explored in this chapter. Moreover, the chapter highlighted the importance of FL, the role of FL in MCPS, and challenges faced by the FL in the MCPS. It also provided a strong foundation for comprehending various FL components and examining their benefits and drawbacks. The implementation strategies and different deployment challenges are discussed, which helped in offering appropriate solutions to the field of hospitals and clinics. Finally, the chapter provided the types and several security threats of FL-based MCPS.

REFERENCES

[1] Xudong, Z., Wei, X., Junfeng, Z., Deterministic event-based physical watermarks against replay attacks in cyber-physical systems, *IEEE Transactions on Industrial Cyber-Physical Systems*, 2, 61–69, 2025.

[2] Lavina, B., Aruchamy, P., An energy-aware software fault detection system based on hierarchical rule approach for enhancing quality of service in internet of things-enabled wireless sensor network, *Transactions on Emerging Telecommunications Technologies*, 35, 1–21, 2024.

[3] Haibin, G., Jain, S., Analysis of replay attacks with countermeasure for state estimation of cyber-physical systems, *IEEE Transactions on Circuits and Systems II: Express Briefs*, 71, 206–210, 2024.

[4] Balraj, L., Aruchamy, P., Sowndarya, K., A novel adaptive Unclonable utility model for secure communication in internet of things environment, *Proceedings in International Conference on Advances in Computing, Control, and Telecommunication Technologies*, 1–6, 2024.

[5] Prasanth, A., Lakshmi, D., Rajesh, D., Balamurugan, B., Technological Advancement in Internet of Medical Things and Blockchain for Personalized Healthcare: Applications and Use Cases, *CRC Press*, 1–218, 2024.

[6] Wenli, D., MengChu, Z., Abdullah, A., A survey of cyber-attacks on cyber physical systems: Recent advances and challenges, *IEEE/CAA Journal of Automatica Sinica*, 9(5), 784–800, 2022.

[7] Divya, S.V., Venkadesh, P., Shiny, K.V., Ninisha, N., Classifying and predicting Covid-19 during pregnancy using RNN and ML techniques, *Proceedings in IEEE International Conference on Social and Sustainable Innovations in Technology and Engineering,* 1–6, 2024.

[8] Insup, L., Oleg S., A medical cyber-physical system for predicting maternal health in developing countries using machine learning, *Healthcare Analytics*, 5, 1–18, 2024.

[9] Simon B., Stefan S., Michael P., Thomas S., Attack detection for medical cyber-physical systems–a systematic literature review, *IEEE Access*, 11, 41796–41815, 2023.

[10] Fadwa, A., Heba, G., Fahd, N., Mesfer, A., Cyber-attack detection in healthcare data using cyber-physical system with optimized algorithm, *Computers and Electrical Engineering*, 108, 1–16, 2023.

[11] Yang, L., Yan, K., Tianyuan, Z., Yanhong, P., Vertical federated learning: Concepts, advances, and challenges, *IEEE Transactions on Knowledge and Data Engineering*, 36, 3615–3634, 2024.

[12] Jingxue, C., Hang, Y., Zhiyuan, L., Min, Z., When federated learning meets privacy-preserving computation, *ACM Computing Surveys*, 56(12), 1–36, 2024.

[13] Yifei, Z., Dun, Z., Jinglong, L., Xinyu, F., A survey of trustworthy federated learning: Issues, solutions, and challenges, *ACM Transactions on Intelligent Systems and Technology*, 15(6), 1–47, 2024.

[14] Tingting, W., Tao, T., Zhen, C., Kai, F., Federated learning-based information leakage risk detection for secure medical internet of things, *ACM Transactions on Internet Technology*, 1–17, 2024.

[15] Yuxia, C., Chen, F., Wenzhuo, S., A blockchain-based federated learning method for smart healthcare, *Computational Intelligence and Neuroscience*, 2021, 1–18, 2021.

[16] Ali, A., Kusum, Y., Yasser, A., Wattana, V., Sandeep, K. Federated learning algorithms to optimize the client and cost selections, *Mathematical Problems in Engineering*, 2022, 1–12, 2022.

[17] Li, L., Yuxi, F., Mike, T., Kuo, Y., A review of applications in federated learning. *Computers & Industrial Engineering*, 149, 1–24, 2020.

[18] Fan, Y., Mohammad, Z., Petr, H., An explainable federated learning and blockchain-based secure credit modeling method, *European Journal of Operational Research*, 317(2), 449–467, 2024.

[19] Chunyan, L., Jiaji, W., Shuihua, W., Yudong, Z., A review of IoT applications in healthcare, *Neurocomputing*, 565, 1–17, 2024.

[20] Li, L., Yuxi, F., Mike, T., Kuo-Y., A review of applications in federated learning, *Computers & Industrial Engineering*, 149, 1–14, 2020.

[21] Bjarne, P., Nico, S., Bert, A., Federated learning in a medical context: A systematic literature review, *ACM Transactions on Internet Technology (TOIT)*, 21(2), 1–31, 2021.

[22] Syreen, B., Moayad, A., Eitaa, A., Nida, M., Federated learning review: Fundamentals, enabling technologies, and future applications, *Information Processing & Management*, 59(6), 1–11, 2022.

[23] Jie, X., Benjamin, S., Chang, S., Peter, W., Jiang, B., Fei, W., Federated learning for healthcare informatics, *Journal of Healthcare Informatics Research*, 5, 1–19, 2021.

[24] Hao, G., Pew-T., Andrea, B., Mingxia, L., Federated learning for medical image analysis: A survey, *Pattern Recognition*, 151, 1–11, 2024.

[25] Jawadur, R., Faisal, A., Nazma, A., Mohammad, H., Challenges, applications and design aspects of federated learning: A survey, *IEEE Access*, 9, 124682–124700, 2021.

[26] Ruijin, W., Jinshan, L., Zhiyang, Z., Xiong, L., Pandi, V., Privacy-preserving federated learning for internet of medical things under edge computing, *IEEE Journal of Biomedical and Health Informatics*, 27(2), 854–865, 2022.

[27] Latif, K., Walid, S., Zhu, H., Ekram, H., Federated learning for internet of things: Recent advances, taxonomy, and open challenges, *IEEE Communications Surveys & Tutorials*, 23(3), 1759–1799, 2021.

[28] Jianping, C., Ximeng, L., Zhiyong, Y., Efficient vertical federated learning method for ridge regression of large-scale samples, *IEEE Transactions on Emerging Topics in Computing*, 11, 511–526, 2023.

[29] Viraaji, M., Reza, M.P., Seyedamin, P., A survey on security and privacy of federated learning, *Future Generation Computer Systems*, 115, 619–640, 2021.

[30] Navod, N., Muneeswari, G., Parthasarathy, V., Fawaz, A., Federated learning for privacy-preserved medical internet of things, *Intelligent Automation & Soft Computing*, 33(1), 157–172, 2022.

10 Use Cases and Scenarios for Federated Learning Adoption in IoT

Anitha Velu, Raghu Ramamoorthy, A. Prasanth, and Ahmed A. Elngar

10.1 INTRODUCTION

Federated learning (FL) refers to a cutting-edge machine learning (ML) methodology, which is essential to the Internet of Things (IoT). This method relies on dispersed learning, in which IoT devices learn collectively without exchanging data with other devices within their on-premises context. FL is a promising paradigm for ML, particularly once it meets IoT devices. It maintains the decentralized nature of the data while enabling numerous devices or nodes to jointly train a common model. Collaborative ML can be made possible with FL, a strategy that avoids data sharing across networks [1]. Industry 4.0, the emerging technology, and IoT present exciting opportunities for FL adoption.

The proliferation of smart devices has led to a rapid expansion of the IoT. Less than 9 billion gadgets were projected to be connected in 2019, but by 2023 that figure had risen to 15 billion, and by 2030 it was predicted to reach 29 billion. Copious data is built by smart devices, and ML systems are frequently used these days to process such data in diverse or crucial application areas, which includes industrial control, environment monitoring, and healthcare. Less than 20 zettabytes (ZB) of data were gathered by devices in 2019; by 2025, it is anticipated that the volume of data related to IoT devices will have increased to 79.4 ZB [2]. Deployment of ML-driven IoT use cases generally requires sending the gathered data to central cloud servers for processing, as seen in Internet of Medical Things [3]. Owing to network constraints, privacy concerns, or application-specific security difficulties, this might not be feasible. Direct data processing at the edge layer is an alternate strategy. Nevertheless, IoT devices with limited resources are not well suited for typical ML techniques [4].

With the help of a growing community, TinyML is a system centered throughout software platforms, hardware, and algorithms that enables the execution of ML algorithms on IoT devices with limited resources. Applications for TinyML can be found in many domains, including Industry 4.0, e-health, transportation networks,

DOI: 10.1201/9781003532323-10

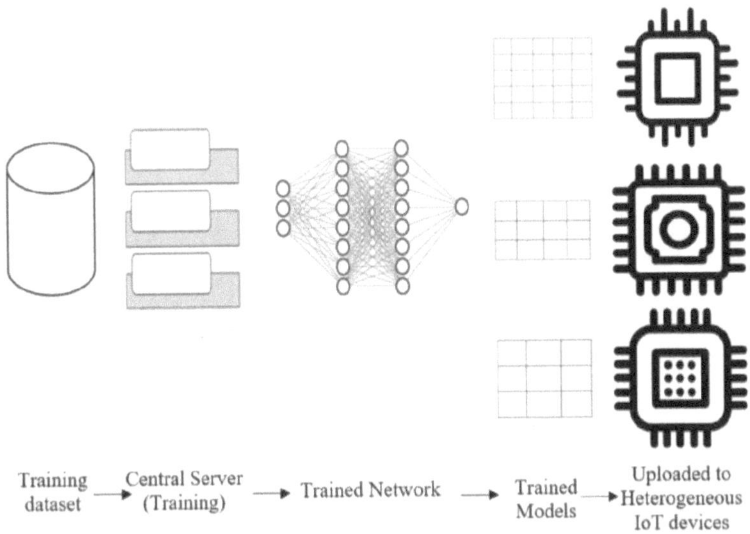

FIGURE 10.1 General model training (MT) process.

and agriculture [5]. On-device model training is not supported by most TinyML implementations available today. After being trained on robust servers, the models are moved to the target devices and activated. As a result, this strategy might not work in industries where privacy is a problem.

Reducing latency and network bandwidth consumption while training the model straight on the devices instead of sending private information via the network is the challenge. The majority of solutions revealed in research works are simulations that demonstrate the practicality of on-device training, which has not been developed enough to use in real-world settings. The traditional model training involves the collection of datasets that have been trained by the central server and converted as an array of weighted networks that can be uploaded to the respective IoT devices, as shown in Figure 10.1.

Development of ML models usually depends on distributed datasets. FL, referred to as collaborative learning (CL), undoubtedly established a recognition for playing a significant role in addressing privacy and security issues that shield complex data or information deriving out of a range of ML-associated threats. This makes FL the perfect option concerning newly developing networks, such as IoT systems [6, 7]. This is especially true given its cutting-edge algorithms, which prioritize their usefulness beyond IoT web even in the existence of appliances with limited resources. However, accomplishment of FL training procedure has been severely hampered by heterogeneous description of existing appliances and training models in intricate IoT web [8]. Continuous efforts have been made by various researchers and scientists to address this problem. However, it remains a significant challenge and is considered nearly unsuitable for direct deployment over IoT networks.

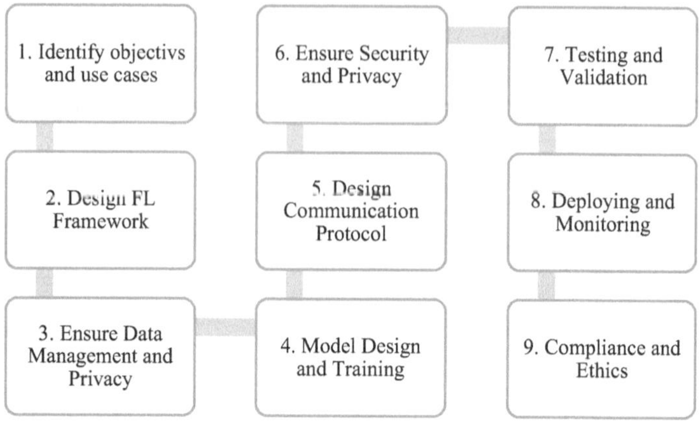

FIGURE 10.2 Key steps to integrate FL with IoT.

10.2 KEY ASPECTS OF ADOPTING FL WITH IOT

FL has been investigated recently as a novel paradigm that lowers latency during model training while maintaining end-user knowledge and data privacy. FL works with a variety of end-user appliances and device applications, even though it is frequently used among central server and edge devices [9]. The way FL should be constructed in this circumstance needs to be carefully adjusted because of the restricted resources of IoT edge devices. Adopting FL with IoT involves several key steps and considerations to effectively integrate and deploy the approach. Figure 10.2 shows a detailed approach to get started with integrating FL with IoT devices.

10.2.1 Identify Objectives and Use Cases

Identification of the specific problem or use case in which FL can benefit the application needs to be defined by the user. Common applications in IoT include predictive maintenance, anomaly detection, personalization, and data-driven optimization. Hence, the user has to define what has to be achieved with the adoption of FL. This might involve improving model accuracy, personalizing user experiences, or reducing data privacy risks [10]. The individual and integrative use of IoT, FL, and information-centric networking have lately been applied in numerous network-related scenarios and have therefore witnessed an increasing interest in the research community.

10.2.2 Design FL Framework

Select a FL framework or library that fits the requirements of the defined objective during the identify objectives process [11]. Popular frameworks include TensorFlow Federated, PySyft (for PyTorch), and International Business Machines Corporation (IBM-FL). The computational and storage capabilities of IoT devices have to be chosen by ensuring that they can handle the local training tasks and communication requirements of FL. Finally, a central server or aggregation point is set to coordinate model training, aggregate updates from devices, and manage exchange of information between the devices and central server.

10.2.3 Ensure Data Management and Privacy

Data management and privacy ensures that the collected data adheres to privacy regulations and best practices on IoT devices [12]. FL enhances privacy, but the developer still needs to manage data securely on each device. Once data is ensured for privacy then it has to be preprocessed and cleaned on the local devices as necessary, so the data used for training is relevant and high quality. By paying user contributions, the developer aims to encourage more clients to take part in crowdsourced FL projects, guaranteeing that the datasets used for model training are diverse and strong.

10.2.4 Model Design and Training

Model design and training involves choosing or designing a ML model that is suitable for the desired use case. Consider the complexity of the model in relation to computing capacity of the IoT appliances. Then develop local training routines for each device and ensure that the devices have the ability to change the model's parameters and train it using local data [13]. Finally, implement an aggregation algorithm on the primary server to combine updating of models deriving out of various devices. Common strategies include federated averaging, secure aggregation, and others tailored to your needs.

10.2.5 Design Communication Protocol

Design communication protocol involves designing a robust communication protocol to handle the swapping of updated models among devices and primary servers. In FL, adoption of an IoT shared model is used; this interaction affects FL training performance and results in a communication overhead. When constructing a FL system, overhead is crucial for choosing the appropriate communication protocol. This includes handling intermittent connectivity and optimizing data transmission. Then implementation of synchronization mechanisms has to be carried out to ensure that all devices and central server are aligned on the trained model state and training progress.

10.2.6 Ensure Security and Privacy

Clients are not required by FL to provide the server access to their sensitive information, unlike centralized ML systems, which save computation to the core server. Obviously, adaptation of FL is not without problems and challenges. From one point of view, client-provided model changes at each training epoch may expose confidential client information. On the other hand, immoral clients may target the model that the server has learnt; these security assaults could contaminate the model or stop it from developing. Ensure security and privacy aids to the aggregation process on the primary server are secure and the updated model from the appliances is protected against tampering and leakage. Next, to improve data protection, federated security features, including homomorphic encryption, safe multiparty computation, and differential privacy, have to be put into practice.

10.2.7 Testing and Validation

In FL, models are refined on the different nodes with different datasets. In order to compare model performances, model validation should be applied independently to each node. Each training cycle in the FL concept can use one of two validation types: (1) Test on global updates, which validates globally modified parameters. It is the process of validating the combined parameters before training a node for the current round. (2) Test on local updates, which validates on locally updated parameters, where validation is applied following the completion of a node's local training for the current round and the local updating of the model parameters. Testing and validation step ensures the developer to test and simulate the proposed model to assess the fulfillment of FL system [14]. This includes testing accuracy and robustness of the model, and its efficiency in real-world scenarios. Then, validation is carried on to validate the performance of the global model using an authenticated dataset or through cross-validation methods to ensure it generalizes well across different devices.

10.2.8 Deployment and Monitoring

Deployment and monitoring insists on deploying the FL system in a live environment, ensuring that the IoT devices are properly configured to contribute to the learning progression [15]. Then, the accomplishment of the FL system has been monitored continuously to track metrics such as model accuracy, training efficiency, and communication costs. Based on monitoring feedback, necessary adjustments can be taken to the model, training procedures, or communication protocols to optimize the system.

10.2.9 Compliance and Ethics

Finally, compliance and ethics aids to ensure that the FL implementation complies with applicable regulations like the California Consumer Privacy Act, the General Data Protection Regulation, or other data protection laws and address ethical considerations related to data use and privacy. Ensure transparency with users about how their data is being used and how their privacy is protected. The developer must examine worldwide patterns and broad guidelines concerning the ethics of FL systems. The primary legal difficulties surrounding FL are pertaining to the fairness and openness of AI systems, privacy and the protection of personal data, intellectual property, legal ownership and rights to the system, and output and governance issues.

By following the abovementioned steps, any developers or researchers can effectively integrate FL with IoT systems, leveraging the benefits of decentralized learning's advantages while tackling the particular difficulties presented by IoT contexts.

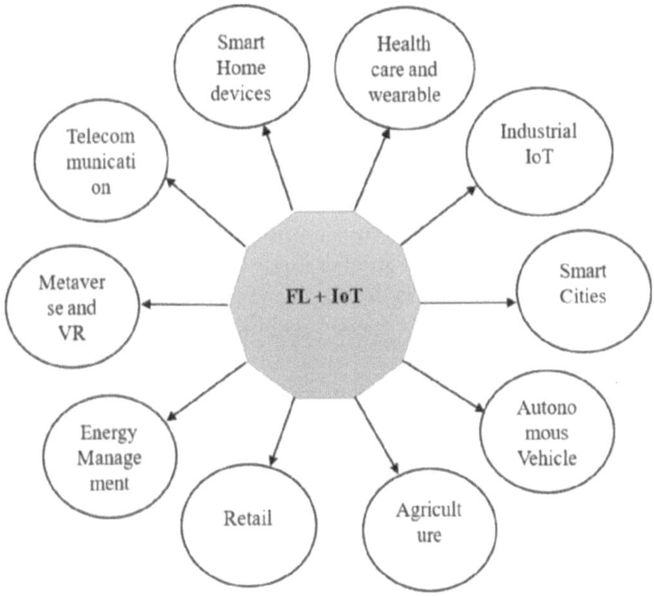

FIGURE 10.3 Use cases and scenarios for FL adoption with IoT.

10.3 REAL-TIME APPLICATIONS OF FL ADOPTION WITH IOT

FL permits collaborative training of models beyond centralized data storage, which has several advantages over IoT applications [16]. Figure 10.3 illustrates some interesting scenarios and use cases for FL adoption in the IoT environment.

10.3.1 Smart Home Devices

IoT devices have become increasingly popular in recent years because they enhance people's comfort and quality of life. The wireless IoT-based smart bulbs, smart cameras, and doorbells have communication capabilities with one another, which has been managed from a distance using mobile phones and microcontrollers. Personalized home automation and digital home systems, such as thermostats, voice assistants, and lighting systems, can benefit from FL to improve personalization. FL enables each device to locally train models based on user behavior and preferences, leading to more accurate and personalized automation while preserving privacy [17].

For instance, a smart thermoregulator can learn individual user choices for temperature settings and adjust automatically without sending personal usage data to the cloud. While creating a home automation architecture with an emphasis on security and privacy, FL is essential since it blends FL with safe data aggregation. As a result, FL has emerged as a crucial technique that can lower the chance of raw data leakage while enhancing on-device speaker verification. FL can be used by smart appliances, including ovens, washing machines, and refrigerators, to enhance their functionality. For example, a smart refrigerator may adjust energy use and food management based on user preferences. FL improves user experience and protects privacy by allowing appliances to learn about and adjust to local user preferences.

10.3.2 Healthcare and Wearable Devices

FL can be used by wearable health devices, including fitness trackers and smartwatches, to create customized fitness and health recommendations. These gadgets gather private medical information, but they also use local model training to make inferences and forecasts like identifying irregular heart rates or anticipating patterns in physical activity. FL protects privacy by certifying the private health data, which stays on device and gradually improves model accuracy and personalization. With the increasing prevalence of IoT devices in people's daily lives, data privacy is becoming a major concern [18]. IoT e-health is one example to show privacy concerns. These days, sophisticated wearables are utilized to trail a patient's level of glucose, blood pressure, and pulse rate. In contrast to other data categories, healthcare data is particularly subtle to the privacy concerns of users and is subject to stringent legal and regulatory restrictions on data sharing. Because of this, investigators and researchers need to use approaches like FL in order to create cutting-edge ML models across a highly regulated and fragmented data ecosystem.

10.3.3 Industrial IoT

Copious sensor data is produced by machinery and equipment in industrial environments. By using FL, predictive maintenance models may be trained locally on each unit to identify possible anomalies or failures before sending raw data from sensor to server hub. This method permits real-time, machine-specific maintenance predictions while minimizing the perils of data leakage and communication expenses. The manufacturing industry is benefiting from various advancements in information technology applications caused by rapid evolution of the Industrial IoT (IIoT).

Idea of Industry 4.0 was put forth in response to the growing importance of IIoT interconnectivity and real-time data availability [19]. Industry 4.0's unparalleled interconnectedness will provide supply chains across a wide range of businesses with more knowledge, control, and data visibility. However, a few practical issues still make Industry 4.0 deployment difficult. First, there's a chance that the volume of data produced by a single firm won't be adequate to fully train a trustworthy model. Second, protecting privacy is crucial as the commercial value of data assembled by IIoT devices is closely linked to it.

10.3.4 Smart Cities

FL helps to enhance traffic surge and environmental monitoring in smart cities. In smart cities, FL can optimize environmental monitoring and traffic control [20]. By leveraging local data and FL, smart city applications can operate efficiently and adaptively while ensuring that sensitive data remains decentralized and secure. For instance, traffic lights and sensors can collaboratively train models to predict traffic patterns and adjust light timings accordingly. Similarly, environmental sensors can aggregate data to predict pollution levels. Generally, sensors and traffic lights working together can train models to anticipate traffic patterns and modify light timings accordingly. In a similar vein, pollution levels can be predicted by combining data from environmental sensors. Applications for smart cities can function effectively and adaptively while guaranteeing that sensitive data is decentralized and secure by utilizing local data and FL.

City administrators can handle things (physical material) in real time and give smart information to users about public safety, traffic, parking, agriculture, transportation, and so on only on account of a variety of IoT gadgets. Smart infrastructures are shifting to computational resources near data locations due to worries about data privacy, which makes the FL architecture appropriate for implementation. The FL-based smart grid technology helps create an intelligent and networked energy exchange system within the city and enables cooperative learning of power consumption trends without revealing specific power traces.

10.3.5 Autonomous Vehicles

As the vehicular IoT continues to progress, regular cars have begun to incorporate autonomous driving technology. Generally, real-time information exchange with a multi-access atmosphere is necessary for a dependable self-driving system [21]. Meanwhile, variations in spatial extension and temporal durations of the vehicle atmosphere necessitate a smart method that grows with the variation in environment. In the case of the standard centralized-over-cloud model, the powerful mechanism desires to transfer a substantial amount of raw data to the server hub that generates possible data breach. Restricted network bandwidth and large-scale data transfer can cause communication overhead, which could hinder the driving system's ability to accurately react to real-time spatial changes.

In order to address the aforesaid issues, FL with vehicular edge computing has become a feasible method for autonomous driving. With FL, each car can respond more sensitively to changes in its immediate environment and requests to send a small amount of information to the cloud. Autonomous vehicles can use FL to enhance navigation systems and safety features. Vehicles can locally train models on data such as road conditions, traffic signals, and driving patterns, sharing model updates without sending raw data. This approach enhances safety and navigation capabilities by aggregating learning from many vehicles while protecting driver and vehicle data.

10.3.6 Agriculture

FL can be used in precision agriculture via IoT sensors on farm equipment and environmental sensors to maximize crop output. By using sensors to train models locally, farmers can improve farming techniques and resource efficiency by analyzing crop health, weather patterns, and soil characteristics. FL maintains the decentralization of sensitive agricultural data while facilitating data-driven, real-time decision-making on farms. Farmers frequently possess sensitive information that they would prefer to keep private. Because FL is an approach that enables several devices to train and share the model without exchanging the data, which is perfect for agricultural applications.

FL applications in agriculture are categorized based on the federation's scalability, architecture, data partitioning techniques, and aggregation algorithms used to create the global model. The federation's scale applies to all FL applications under discussion, with the exception of one cross-device. Researchers and developers go over many FL frameworks as well as the numerous trained models like deep learning and ML for various agricultural domains [22]. Using its own data, FL enables each farm to train the model locally without depending on other sources. Then, beyond exchanging the source data, local models are combined into a global model, protecting privacy and cutting down on data transfer expenses. FL can help in developing customized precision agriculture models for different regions or individual farms.

10.3.7 Retail

E-commerce 5.0 has been highlighted as requiring retail data analysis as a critical component. ML and complex data analysis are two examples of the cutting-edge technologies that have transformed retail data analysis due to the explosive growth of information and communication technology. When creating an application, an FL-based trade data analysis context assists in addressing security and privacy concerns [23]. The FL framework enables the formation of trade data analysis alliances through several foundations that require direct exchange of consumer data, as long as the data is kept locally to protect customer privacy.

FL can be used by retail IoT equipment, like smart shelves and in-store cameras, to improve customer experience and inventory management. Smart shelves have the ability to train models for predicting stock levels and optimizing refilling timetables, all without transferring intricate transaction data to cloud servers. This leverages the collective intelligence of multiple retail locations while preserving privacy and increasing operational efficiency. In summary, the FL framework opens the door to more efficient and safe data-driven resolution or conclusion in the retail sector by providing a viable method for retail data analysis that protects privacy in the age of E-commerce 5.0.

10.3.8 Energy Management

Smart grids can benefit from the application of FL to optimize energy distribution and consumption. Without consolidating sensitive energy consumption data, smart meters and grid sensors can locally train models to forecast energy usage patterns and enhance load balancing. This method lowers data transport costs, protects user privacy, and improves the efficiency of energy management systems.

Generalization of the model is a crucial performance indicator. When opposed to a model that just performs well on its original training dataset, a local model trained on a client's dataset that generalizes well to several more customers may be seen superior in FL [24]. FL is most expected to execute well on other classical ML learning approaches and previously unreported data. The FL approach outperforms

TABLE 10.1
Comparison of Different Training Approaches

	Accuracy (in %)			
	MNIST Dataset		CIFAR-10 Dataset	
Model Training Approach	**TPR (%)**	**TNR (%)**	**FPR (%)**	**FNR (%)**
Centralized Machine Learning	65	73	54	62
Distributed Machine Learning	72	78	67	72
Federated Learning	92	97	86	84

classical training approaches like centralized ML and distributed ML, whose models have been trained on two commonly used open-source datasets, MNIST and CIFAR-10. The evaluation is conducted for two scenarios: (1) 50 participants with 20% participation ratio for 100 communication rounds and (2) same parameters for 200 communication rounds. The resultant parameters are illustrated in Table 10.1 where FL achieves greatest accuracy of about 92% and 97% for scenarios 1 and 2 in MNIST dataset and 86% and 84% accuracy for scenarios 1 and 2 in CIFAR-10 dataset.

A next hypothetical version of the internet known as the "metaverse", which offers completely connected, alluring, and captivating online 3D experiences virtually via traditional computers and virtual reality and augmented reality gadgets. Users in the metaverse have control over the avatars of the user and communicate with other users or virtual items. Digital twins, said to be virtual replicas of the physical world that reflect its current condition, are one of the primary components of the metaverse. The data or information collected from IoT gadgets connects the virtual and real worlds. FL presents a viable way to improve global performance by facilitating edge-server collaboration while also strengthening the security and privacy of metaverse.

For instance, local devices can be educated with eye tracking or motion tracking data gathered from millions of wearable users, and the data can then be pooled through an FL server. Users can thereby take advantage of metaverse services without jeopardizing their privacy. Federated learning for metaverse (FL4M) is identified to be an effective tool. This is due to FL's ability to let edge devices use their own data, processing power, and model-building capability to take part in training tasks locally. By implementing FL into the metaverse, users' data privacy is safeguarded and servers require less processing power and memory. There has been numerous research on FL and the metaverse up to this point.

10.3.9 Telecommunications

With the successful global deployment of 5G networks, industry and academia are now examining 6G technology for experiencing better communication associated with consumer applications. In order to support ML deployed data-driven approaches on immensely scalable and diverse network applications, 6G will be built on a pervasive FL platform. The traditional ML technique entails centralizing train data in data centers so that data inference and analysis may be performed using centralized ML algorithms [25]. Due to privacy concerns and insufficient communication capabilities in wireless networks, it is sometimes inconvenient or impractical for the devices to transfer information to the designated server for data inference and analysis.

Nevertheless, it is often impractical or inconvenient for any things or gadgets to agree to the transmission of data to the restricted server, due to privacy concerns and limitations in wireless network transmission bandwidth. FL, which may address these problems, allows the electronic gadgets to train a standard and useful model with requirement of data flow and interchange. FL is a useful tool for telecommunication networks to improve service quality and network performance. Together, base stations and user devices may train models to anticipate network traffic patterns and maximize the distribution of resources. FL enhances user experience and network performance without necessitating centralized data collecting.

By leveraging FL, these IoT scenarios can benefit from enhanced privacy, reduced data transfer costs, and more personalized and efficient systems. Each use case demonstrates how FL approach helps in applying itself to solve specific challenges by addressing the constraints and requirements of IoT environments.

10.4 CHALLENGES IN ADOPTING FL WITH IOT

The prospect of FL adoption with IoT discovers seven major challenges that prevent FL from being enabled to adopt with numerous IoT devices [26]. The identified challenges (Figure 10.4) prevent FL from reaching its full potential in the applications discussed above. These issues stem from (1) limited on-device resources of IoT devices, (2) limited bandwidth of network at the edge, (3) the inconsistent availability and connectivity that frequently occur in physical-world sceneries, (4) the variability and variety of IoT devices available with respect to their resources, (5) historical dynamics and continual updating of models following the distribution of devices, (6) the ways to secure client information and defend against adversarial attacks, and (7) the absence of community standards and system development tools. Researchers and developers who are working with integration of FL with heterogeneous IoT devices must overcome these challenges.

10.4.1 Limited On-Device Resources

The constrained resources of IoT devices impede a substantial obstacle to the implementation of FL at the edge of the network. Deep neural networks, in particular, are known to be computationally demanding ML models, where a precise hardware may result in the least effective training on edge devices. Therefore, developing dedicated hardware for edge ML applications is a feasible approach to accelerate training and inference procedures while using less energy than conventional CPUs. Edge devices have constrained memory for data access and storage in addition to low computational power. Researchers primarily concentrate on two research paths to increase the applicability of models for the edge environment. One is creating models and methods that are light and compatible with hardware, and compressing the existing

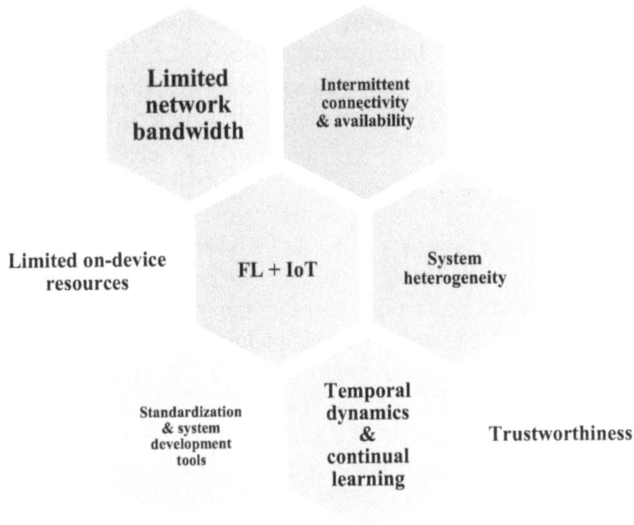

FIGURE 10.4 Challenges in adopting FL with IoT.

or state-of-art models to create stripper, minor models that require less compute and energy.

10.4.2 Limited Network Bandwidth

An information exchange bottleneck is perceived as a major challenge for an IoT environment based on FL. The majority of IoT devices now use wireless networks for communication, which have far less bandwidth than wired networks in data centers. For instance, regardless of the number of connected devices, the total networking bandwidth for the IoT system remains constant in a smart home setting. As more devices are added to the system, communication issues arise when clients allocate resources differently. In addition to causing inefficient client-server connection, the constrained network capacity also leads to the emergence of straggler clients, which neglect to provide the server with their local updates between communication cycles. In the large-scale IoT scenario, they both act as the bottlenecks for the FL deployment performance.

10.4.3 Intermittent Connectivity and Availability

In addition to the bandwidth constraint previously mentioned, the sporadic linking of IoT gadgets indicates an unbalanced network connectivity that causes the gadget to disconnect from the system mid-training cycle. Dropout issues combined with erratic device availability and connectivity, particularly in large-scale IoT systems, will provide a significant challenge to the capacity of FL context to effectively schedule and manage clients. The majority of FL trainings currently in use are predicated on the server's synchronous update, which suggests that the model aggregation won't begin until the server has received the data from the slowest client. In order to overcome this difficulty, several researchers suggested an asynchronous aggregation approach that uses coding theory to withstand the FL system's stragglers.

10.4.4 System Heterogeneity

Clients under the FL framework have different hardware and software system metrics in cross-device scenarios. The learning activities are carried out under different operating systems and software application programming interfaces (APIs) using a variety of gadgets through varied hardware designs, including distinct device suppliers. To train the local models, clients can select from a variety of DL foundations, including PyTorch, TensorFlow, and Caffe which will produce a variety of model formats for aggregation. An IoT FL framework should allow data and computation load to be gracefully adapted across various devices according to their resource availability. In order to overcome this difficulty, researchers and developers must identify the heterogeneity and necessary modifications that have to be made prior to the training session, hence both the quality and speed of the training will increase.

10.4.5 Trustworthiness

In real-world deployments, adversaries look to carry out Distributed Denial of Service (DDoS), phishing, and identity theft to find IoT devices for appealing targets. The latent traffic capacity of IoT-based DDoS assaults is growing as a consequence of propagation of IoT web. This has been demonstrated by the Mirai botnet attack, which leverages compromised cameras and home routers. Internet-based attacks have brought attention to the necessity of IoT hazard calculation and security, for instance, in industries like healthcare.

Numerous IoT devices lag in computational power necessary to apply security fixes, despite the fact that doing so would readily defend against these threats. Furthermore, it can be challenging to determine if a new participant in a cross-device system is malicious before it joins the system. Therefore, to avoid ruining the model training with limited resources, the IoT system needs to be able to detect malicious or faulty IoT devices. Using a trivial security protocol to recognize malicious and malfunctioning gadgets in the IoT is said to be a promising solution to this problem. FL can provide an alternate method of IoT cybersecurity because of its distributed architecture, which protects the structure from harmful outbreaks on IoT gadgets.

10.4.6 Temporal and Continual Learning

By definition, IoT distinguished appliances will always be gathering fresh information, which is utilized to upgrade the training model for ongoing or lifelong learning. For IoT devices with limited resources, the ongoing model update training presents a new difficulty in order to maintain services adapting to freshly collected data. In particular, the cache resources of middle-of-the-road IoT devices are insufficient to support both model inference and training because they have limited memory. Moreover, one of the most important issues with continuous learning, catastrophic forgetting may be made worse by a memory deficit that prevents one from retaining previously acquired information. Researchers believe that in order to overcome this difficulty, a lightweight ML engine is required for lowering the storage capacity for on-device training.

10.4.7 Standardization and System Development Tools

While integrating FL systems on IoT networks, researchers have a lot of things to consider. It is necessary to take into account factors like various communication APIs, dataflow models, network setups, and device attributes. FL for IoT is still in its infancy and hasn't been well standardized or implemented. As a result, standardization-related research and development may hasten the widespread installation of FL approach on IoT mesh and foster the atmosphere that is conducive to content sharing. Furthermore, given the intricacy of FL, enterprises and researchers must continue to improve on already-existing FL benchmarking and development tools like FedML, TensorFlow Federated, and PySyft to support some scenarios of IoT proposals.

10.5 ADVANTAGES OF FEDERATED LEARNING IN IOT

FL enables resident model training on gadgets, which helps to divide the quantity of data that must be transmitted across the network. It also makes it possible to train models that can be used on appliances with minimal resources, like smart phones or IoT sensors. Advantages of FL adoption with IoT are elaborated as follows:

Data Privacy: FL reduces the amount of private information that is accessible to outside parties, which is one of its benefits. Personal data stays on individual user devices and not sent to the primary server because the model training occurs locally on individual devices. Additionally, FL uses privacy-preserving methods like differential privacy and encryption to further protect user data while it is being trained. FL reduces the risks of data breaches and unauthorized access by eliminating centralized storage and maintaining data localization. Furthermore, FL minimizes the likelihood of data exposure to outside parties by guaranteeing that user data stays under the control of local servers or individual devices. Sensitive information is kept confidential and user privacy is improved by this decentralized method of data processing and storage. IoT devices often handle sensitive personal data. FL ensures that raw data never leaves the device, reducing privacy concerns since only the updated model, but not actual data is shared with the primary server.

Reduced Latency: By training models locally on the device, FL can minimize the need for data transfer that results in better benefits by providing low-bandwidth or high-latency networks. Minimizing latency in a wireless FL system with heterogeneous processing capabilities, in which the quantity of local model updates is carried out by various edge devices varies with each communication round. Considering both the communication and compute latency in the entire FL process, a total latency minimization is done with probabilistic device selection. In order to mitigate the straggler effect and address this simplified problem, optimization approach takes into account both communication and computation resource allocation. Allocation of processing capacity and bandwidth are both convex and amenable to optimal solutions in closed form.

Scalability: More data leads to stronger models in data science. However, training a model with large datasets might take a long time if all the data is stored on a single server or cloud. FL speeds up AI and IoT deployment models and increases operational scalability by enabling the storage of datasets in numerous locations and training models on each piece of data at the same time. FL is found to be appropriate for situations with an enormous number of devices connected to the environment. It can efficiently scale across numerous IoT devices without requiring centralized data collection. Additionally, FL minimizes the amount of data that must be sent to a network and analyzed in the cloud by orders of magnitude because it trains the model at the edge and just moves the trained weights. In this manner, it lowers computer and storage expenses and does away with the need for expensive data transfers.

Personalization: Model personalization is a popular method for enhancing model performance in FL when data heterogeneity (data that is not uniformly dispersed across data silos) occurs in the model. Local models can be tailored to specific devices or user needs, offering more personalized and relevant predictions or actions. In fact, current benchmarks indicate that fitting distinct local models on the federated data is already competitive and that practical federated datasets may be extremely varied. A distinct trade-off between the utility costs of privacy and data heterogeneity may arise when taking into account model personalization strategies under silo-specific example-level privacy.

Energy Efficiency: FL can reduce the energy consumption associated with data transfer, as only model parameters are communicated rather than raw data. A numerous intelligent reflecting surface (IRS)-assisted multiuser communication system can be taken into deliberation as a means to minimize the overall energy usage during the training phase. Federated deep learning insists to identify the best reflection configurations of each IRS in parallel. FL empowers training large fleets and several clients together without sharing the training datasets. By safeguarding data privacy, FL could be able to solve the dearth of data sharing in the renewable energy sector, which is preventing innovation, research, and development.

10.6 CONCLUSION

FL has gained an ability to significantly advance the capabilities of IoT systems by addressing data privacy concerns and enhancing personalization while dealing with the challenges of heterogeneity and communication costs. As the technology matures, one can expect its adoption to grow, leading to more efficient and intelligent IoT ecosystems. FL speeds up AI and IoT deployment models and increases operational scalability by enabling datasets to be stored in numerous locations and training models on each piece of data concurrently. Furthermore, FL significantly divides the quantity of data that must be transmitted across the network and analyzed in the cloud since it trains the model at the edge and just moves the trained weights. FL also lowers computer and storage expenses and does away with the need for expensive data transfer. FL trains models without sending the raw data because of its collaborative nature, which treats every IoT gadget at the edge as a distinct client. This implies that each IoT device only collects the data required to finish its mission during the FL process. FL minimizes the danger of personal data leakage, ensures private raw data, and guarantees operational security by storing it on local memory of the device by just conveying model updates to the primary server.

REFERENCES

[1] Safri, H., Kandi, M.M., Miloudi, Y., Bortolaso, C., Trystram, D., Desprez, F., Towards developing a global federated learning platform for IoT, *IEEE 42nd International Conference on Distributed Computing Systems (ICDCS)*, Italy, 2022, 1312–1315.

[2] Vailshery L.S., *Number of IoT Connected Devices Worldwide 2019–2021*, Statista, 2024.

[3] Ahmed, S.F., Alam, M.S.B., Afrin, S., Rafa, S.J., Rafa, N., Gandomi, A.H., Insights into Internet of Medical Things (IoMT): Data fusion, security issues and potential solutions, *Information Fusion*, 102, 32–46, 2024.

[4] Aruchamy, P., Lavina, B., Devi, S. An energy-aware link fault detection and recovery scheme for QoS enhancement in Internet of Things-enabled wireless sensor network, *Computers and Electrical Engineering*, 123, 1–17, 2025.

[5] Sanchez-Iborra, R., Skarmeta, A.F., Tiny ML-enabled frugal smart objects: Challenges and opportunities, *IEEE Circuits and Systems Magazine*, 20, 4–18, 2020.

[6] Velu, A., Ramamoorthy, R., Saravana Kumar, E., Shruthi, E., IoT-enabled smart farming: A controlled environment agriculture application, *International Conference on Sustainable Communication Networks and Application (ICSCNA)*, India, 2023, 292–298.

[7] Velu, A., Ramamoorthy, R., Manasa, S.M., Navulkumar, D, An energy efficient IoT based smart street lighting using low-cost SOC, *International Conference on Electronics, Computing, Communication and Control Technology (ICECCC)*, India, 2024, 1–6.

[8] Yaacoub, J.A., Noura, H.N., Salman, O., Security of federated learning with IoT systems: Issues, limitations, challenges, and solutions, *Internet of Things and Cyber-Physical Systems*, 3, 155–179, 2023.

[9] Graziele Ferreira, S.L., Djamel F.H., Patricia Takako E.S., Resource optimizing federated learning for use with IoT: A systematic review, *Journal of Parallel and Distributed Computing*, 175, 92–108, 2023.

[10] Rahman, A., Hasan, K., Kundu, D., Islam, M.J., Debnath, T., Band, S.S., Kumar, N., On the ICN-IoT with federated learning integration of communication: Concepts, security-privacy issues, applications, and future perspectives, *Future Generation Computer Systems*, 138, 61–88, 2023.

[11] Kholod, I., Yanaki, E., Fomichev, D., Shalugin, E., Novikova, E., Filippov, E., Nordlund, M., Open-source federated learning frameworks for IoT: A comparative review and analysis, *Sensors*, 21, 167–172, 2021.

[12] Lavina, B., Aruchamy, P., Sowndarya, L., Kuntavai, T., A lightweight blockchain scheme for secure data communication in internet of things-enabled wireless sensor network, *Proceedings in International Conference on Smart Systems for applications in Electrical Sciences*, Siddaganga Institute of Technology, Tumakuru, Karnataka, India, 1–6, 2024.

[13] Colakovic, A., IoT systems modeling and performance evaluation, *Computer Science Review*, 50, 162–178, 2023.

[14] Wen, J., Zhang, Z., Lan, Y., A survey on federated learning: challenges and applications, *International Journal of Machine Learning and Cybernetics*, 14, 513–535, 2023.

[15] Ficco, M., Guerriero, A., Milite, E., Palmieri, F., Pietrantuono, F., Russo, S., Federated learning for IoT devices: Enhancing TinyML with on-board training, *Information Fusion*, 104, 145–161, 2024.

[16] Yadav, S.P., Bhati, B.S., Mahato, D.P., Kumar, S., Federated learning for IoT applications, *EAI/Springer Innovations in Communication and Computing*, 1, 1–17, 2022.

[17] Aivodji, U.M., Gambs, S., Martin, A., IOTFLA: A secured and privacy-preserving smart home architecture implementing federated learning, *IEEE Security and Privacy Workshops (SPW)*, San Francisco, 2019, 25–37.

[18] Prasanth, A., Lakshmi, D., Dhanaraj, R. K., *Technological Advancement in Internet of Medical Things and Blockchain for Personalized Healthcare: Applications and Use Cases*, CRC Press, 1–290, 2024.

[19] Farahani, B., Monsefi, A. K., Smart and collaborative industrial IoT: A federated learning and data space approach, *Digital Communications and Networks*, 9, 436–447, 2023.

[20] Ghadi, Y. Y., Mazhar, T., Shah, S. F. A., Haq, I., Ahmad, W., Ouahada, K., Hamam, H., Integration of federated learning with IoT for smart cities applications, challenges, and solutions. *PeerJ. Computer Science*, 9, 57–67, 2023.

[21] Zhang, H., Bosch, J., Olsson, H. H., End-to-end federated learning for autonomous driving vehicles, *International Joint Conference on Neural Networks (IJCNN)*, China, 2021, 18–29.

[22] Yogeshwari, M., Prasanth, A., Smart dust technology for monitoring and control systems in smart agriculture and crop surveillance systems, *Computer Vision in Smart Agriculture and Crop Management,* 123–137, 2025.

[23] Chongder, S., FEDRETAIL: A framework for distributed retail data analysis and learning toward E-commerce 5.0, *International Journal of Advanced Research in Science Communication and Technology*, 2024. https://doi.org/10.21203/rs.3.rs-4100205/v2.

[24] Grataloup, A., Jonas, S., Meyer, A., A review of federated learning in renewable energy applications: Potential, challenges, and future directions, *Energy and* AI, 17, 132–146, 2024.

[25] Song, Y., Xie, Y., Zhang, H., Liang, Y., Ye, X., Yang, A., Ouyang, Y., Federated learning application on telecommunication-joint healthcare recommendation. *IEEE 21st International Conference on Communication Technology (ICCT)*, China, 2021, 42–49.

[26] Shahzad, K., Shakeel, K., Factors influencing the adoption of Internet of Things (IoT) in university libraries: A systematic literature review (SLR) Abid Iqbal. *The Electronic Library*, 42, 121–134, 2024.

Index

C